JOURNAL OF MULTI BUSINESS MODEL INNOVATION AND TECHNOLOGY

Volume 1 No. 2 February 2013

PETER LINDGREN, MORTEN KAMØE SØNDERGAARD, MARK NELSON, and B.J. FOGG / Persuasive Business Models   71-100

KRISTIAN BONDE PEDERSEN, KRISTOFFER ROSE SVARRE, DMITRIJ SLEPNIOV, and PETER LINDGREN / Global Business Model – a step into a liquid business model   101-114

PETER LINDGREN and OLE HORN RASMUSSSEN / Conceptualizing Strategic Business Model Innovation Leadership for Business Survival and Business Model Innovation Excellence   115-134

Published, sold and distributed by:
River Publishers
P.O. Box 1657
Algade 42
9000 Aalborg
Denmark

Tel.: +45369953197
www.riverpublishers.com

Journal of Multi Business Model Innovation and Technology is published three times a year. Publication programme, 2012–2013: Volume 1 (3 issues)

ISSN#: 2245-8832
ISBN# 978-87-92982-41-4
All rights reserved © 2013 River Publishers

No part of this work may be reproduced, stored in a retrieval system, or transmitted in any form or by any means, electronic, mechanical, photocopying, microfilming, recording or otherwise, without prior written permission from the Publisher.

JOURNAL OF MULTI BUSINESS MODEL INNOVATION AND TECHNOLOGY

Editor-in-Chief: Dr. Peter Lindgren, Aalborg University, Denmark

Steering Board
Marcus Eisenhard, Fraunhofer , Germany
Giovanni Schiuma, Universitata Basilicata, Italy
Rasmus Blom, Grundfos Management, Denmark
Joe De Salvo, General Electric, New York, USA
David Nordfors, Stanford University, USA
Debra Amidon, Entovation, Boston, USA

Editorial Board
Professor Giovanni Schiuma, Universitata Basilicata, Italy
Associate Professor Jeppe Gustavson, Aalborg University, Denmark
Associate Professor Robin Tegeland, University of Stockholm, Sweden
Professor Demetres Kouvatsos, University of Bradford, UK
Professor Gregory Yovanof, AIT, Greece
Professor Jeffrey Snape, Harvard University, USA
Professor David Nordfors, Stanford University, USA

Aim of the journal
Provides an in-depth and holistic view of Multi Business Model and technology Innovation from practical to theoretical aspects covering topics that are equally valuable for practitioners as well as academia - also those new in the field.

Scope of the journal
The journal covers Multi Business Model and technology innovation issues and solutions thereof. As Business has moved towards a world of multi business models, issues in modeling business and business models will be published. The publication takes a holistic, strategical, network based and global view to the Multi business approach. Some example topics are: Multi Business Model Innovation and technology in Health Care sector, Multi Business Model Innovation Leadership and management in SME´s, Cloud based Multi Business Model Innovation, Business Model Eco systems, Open and Closed Business Models

PERSUASIVE BUSINESS MODELS

Peter Lindgren [1], Morten Karnøe Søndergaard [2], Mark Nelson [3], and B. J. Fogg [3]

[1] Department of Mechanical and Manufacturing Engineering, Denmark
[2] Department of Learning and Philosophy Aalborg University, Denmark
[3] Center for Persuasive Technology, Stanford University, California, USA

Received: 20 February 2013; Accepted: 28 February 2013

Abstract

This paper will look into how persuasive technologies can be applied into business models in specific business models that apply change in social, society and ecological behavior. We look at which of these business models succeed and which do not and address what abilities and triggers we might increase to change not only attitude but also behavior, when applying business models based on positive ecological, social and society behavior. Focusing on a well-established step-wise persuasion innovation process we start out by identifying past experiences with these types of business model and micro-payments introduced to support change in ecological, social and society behavior.

The pivotal question being, how we get people to make a micro-payment when supporting Business Model´s applying positivesocial, society and ecological behavior. Finally we will brush-upon, how the suggested persuasive business models in a business model context, might also be adapted into other business models and lines of behavior with similar feats. In context of the latter, it is suggested, that there is a potential new business model eco-system on the rise in this field. In order to counteract these effects and create leverage the idea of performing a bundle of business models – a multi business model approach - with persuasive technology embedded is introduced.

Keywords: Persuasive business models, Multi business models Business Models and micro-payments, Business Models and social capital, business model eco-system, persuasive technology

Journal of Multi Business Model Innovation and Technology, Vol 1, 71–100.
© 2013 River Publishers. All rights reserved.

1. INTRODUCTION - WHY BUSINESS MODEL INNOVATE?

In the past ten years the number of persuasive technologies in our everyday life, have increased many-fold. The study of these technologies, and how they affect our lives and routines, has become a study of great interest (Fogg 2012). Industry and public players alike are keenly devoting themselves to understanding how different persuasive technologies might be designed, so that desirable behaviors are obtained. Consider, for example, how your GPS kindly warns you not to use while driving. This is good behavior - to you, and your fellow road users. Or think of, how fitness APPs (you surely have one) might help squeeze that little extra effort out of you, not to speak of, how online social networks can generate vast real life changes, ranging from romantic relationships (memories could bring a smile to your lips) to overthrowing corrupt governments (remember the Arab spring). In other words, tremendous impact can be obtained, if the desired message is transmitted, accepted and carried out as a new behavioral pattern. However, getting this right from the beginning is far from easy and forefront business model researchers and practitioners in the area are therefore very much devoting themselves to the design question.

In this matter we shall draw on B.J. Fogg's eight step design process to creating persuasive behavior (Fogg, 2009) and relate this to Business Models and Business Model innovation focusing on changing social, society and ecological behavior. Simply listed the eight steps involved in Fogg's model are:

1) choose a simple behavior to target

2) choose a receptive audience

3) find what is preventing the target behavior

4) choose an appropriate technology channel

5) find relevant examples of persuasive technology

6) imitate successful examples

7) test and iterate quickly

8) expand on success.

As, explained by Fogg the eight steps are not intended to be used as a rigid formula, and corners are ment to be cut. For instance, there is very little flexibility concerning the technology channel, as there is already in most cases designed a fixed system to use in the use cases we focused on and studied. On the other hand there is a lot of flexibility concerning target behavior and audience.

Digging into the details, in this paper we will look into how persuasive technologies can be applied to change behavior related to business models. Working with Fogg's approach we accept that climbing Mount Everest is done one step at the time, which means that our challenge is divided into a number of smaller steps. The very first is to study the "simple behavior" the business models target to change. This targeting tie in with the technology at hand, which can be a piece of integrated software allowing people to make micro payments when e.g. pushing a "green" button in our printer console (Karnøe et al, 2012) or a "charity button" in a bottle automat (Charity Button Case appendix 1). This in effect means that when handing in used bottles for recycling we are all able to make micro-payments that will be allocated to saving wild animals, re-foresting or/and other social and society programs, so that the overarching and global social, society and ecological consequences of mans act be leveraged. A piece of integrated technology in a recycling bottle automat, which allow us to make micro-donation that will be allocated to help wellbeing projects in WWF – (www.wwf.dk) and Red Barnet – (www.savethechild.dk). About 1100 bottle automats have this persuasive technology installed in their BM, which make it possible to customers to donate the whole or part of the recycling bottle money by pushing "the charity button".

2. DESIGN/METHODOLOGY/APPROACH

The behavior we will target, is simply, how we make people "push the button"? How can we persuade someone to adopt a certain behavior and "buy" a Business Model? And, when doing so, what will the most perspective persuasive business models look like? And how can we build in the persuasive technology in the business models. To come to grips with this challenge, we will start out by looking at previous successes and failures in relations to business models and micro-payments related to social, society and environmental issues. We will then evaluate, and try to identify the biggest barriers to the target behavior and persuasive business models as such. We will address the issues of ability and hot triggers and then try to identify what the optimum set-up in terms of these persuasive business models would be. The framework of this article is then in the "save the world" category and scale, but our goal is much simpler, as we hope to generate a better understanding of what might persuade people in their everyday life to "push the button" and create environmental, social and/or societal leverage.

3. BUSINESS CASES - EXAMPLES TO LEARN FROM

Addressing the challenge of making people do micro-payments in relations to such business models have been brushed upon before and in this chapter we focus on putting some examples to the front. We will dig into the detail of each case, and try to identify, what worked and what was not successful.

Table 1 Purpose of persuasive business models related to ecological, social and society behavior

Purpose and scopes of persuasive business models
Life aid – helping and supporting people to get out of poverty, illness, clean water
Health care – helping people to better healthcare, fighting diseases, getting medicine
Education – helping and supporting people to education
Investment in things – helping people to clean water equipment, energy- and lighting equipment, community and cultural building and equipment, schools
Business development – helping entrepreneurs or small enterprises to develop business
Environmental protection – Carbon reduction, energy efficiency and renewable energy
Society development – supporting people to re build, build up communities, infrastructure

Persuasive Business Models with Micro payment and Micro financing is carried out in many forms. They are also related to many different purpose and scopes.

The list is properly larger however through our case study we found that some overall groups of persuasive business models exist. Some business models address some very specific areas and are very clear in objectives and structure. Others are more blurred in their focus and give donators more opportunities – but also more difficulties to understand and track to their donators. Some business models projects we therefore had to place in more than one group of persuasive business models. In the following we take out one and in some cases two use case to exemplify each group. For a more detailed use case description this can be seen in appendix 1.

3.1. Group 1: Life and healthcare issues (Simpa, CBM, TCE, KB, PCBD, RAD, CSA, HSJD, SOS)

These BM´s have their main focus on life aid due to hunger, war, earthquake, decease e.g. or generally focus at health care as preventing or diminishing AIDS, Cancer e.g. In this group we also placed business models which address more indirectly life and healthcare purpose by preventing people from using unhealthy energy sources.

The objective of these BM´s e.g. the TCE use case, as one representing this group, is to reach total control with AIDS in Southern Africa. TCE builds upon the belief that people by them self can win the battle of AIDS and prevent the HIV infection, while all people can participate via micro donation in "this battle". TCE is micro financed via sales of a special TCE newspaper either by paper or electronically. The overhead of the business models is used for the project

3.2. Group 2: Projects and things (Schools, buildings, issues (Simpa)

These BM´s have their main focus on supporting projects, financing things, co-financing projects and things.

Simpa Networks saw that worldwide, approximately 1.6 billion people have no access to electricity and another 1 billion have extremely unreliable access. Without ready access to electricity, the poor depend on kerosene lanterns and battery-powered flashlights for light. Kerosene lanterns are dangerous, dirty, and dim. Worse, they are very expensive to operate. And yet, in most markets, kerosene lanterns are the preferred lighting system. For a person with little or no savings, no access to formal credit, and low and uncertain income, the selection of kerosene lighting is eminently rational. Simpa developed a solar based micro payment light system and business model, where the consumer pays via mobile connection pr. Use of the system.

3.3. Group 3: Community, social responsibility and society issues (Simpa, CCA, KB, SOS)

These BM´s have their main focus on supporting initiatives related to community development, personal and social support to people suffering from decease or supporting their relative, helping poor children to education.

The Church Cross Army collects "stuff" that people do not any longer want to keep and then sell these items in special CCA shops. The money collect is then donated for certain social and community projects or send to rural areas for help and support. The products are sold by volunteers and then small money donation from each product is then used for donation. The system has a kind of double donation purpose as people come with their stuff and give it to CCA and then other people can come and buy "the stuff" for a low price. The system further has a recycling impact on society as old "stuff" is reused.

3.4. *Group 4: Environmental and nature – (BA, CBM)*

These BM´s have their focus at environmental issues and protection of nature as e.g. reduction of carbon, water protection, nature protection, protection of places of importance to human beings.

By pushing a "charity button" at the bottle automat (Appendix x) at the largest retail chain in Denmark Coop (www.coop.dk) it is possible to donate bottle recycling money to wellbeing projects in WWF – (www.wwf.dk) and Red Barnet – (www.savethechild.dk). About 1100 bottle automats have this extra button in Denmark. It is possible to donate the whole or part of the recycling bottle money. This is done by pushing "the charity button" first, and when the customer reaches the amount they want to donate then they push the charity button. Afterwards the rest of the bottles and cans are delivered and the pay button is pushed and a cash ticket can then be changed to cash at the cash register.

British Airways' is another example in this group as they decided in 2008 to establish a "unique One Destination Carbon Fund" where they took customers flight ticket donations and used them to support energy efficiency and renewable energy projects in communities in UK. BA was targeting parts of UK that BA thought needed the most help to improve their economic and social well-being. The funds were managed a not-for-profit charity fund, using the UK Carbon Reporting Framework as quality insurance. The donation funded a range of projects from helping install solar hot water in community swimming pools, small scale wind turbines for schools or energy efficiency measures in social housing.

3.5. *Group 5: Business Development – (IBS, SBT, GB)*

These BM´s have their focus at business development and/or making it possible for people in general, poor people to establish new business or businesses based on social capital and social benefits.

Cell phone banking have revolutionizes financial services for the poor. A woman manages e.g. a village from one cell phone for a project in rural Bangladesh. Another woman manage equally in rural area in South Africa food supply and banking. SKS, Indias biggest microfinance institution (MFI) expects that an infusion of private capital will spur even greater growth in credit to India's rural poor, where nearly estimated 27m of whom are already microfinance clients (MFI 2010). Banks and cell phone companies are taking advantage this expansion of cell phone use in developing economies to extend financial services to roughly 2 billion people, who use cell phones but lack bank accounts.

3.6. *Group 6: Entertainment (Flatr, RAD, CSA)*

These BM´s have a focus on entertainment as driver to sponsoring, donating and micro financing of a business as such, a project or organizations activities.

The latest years there has grown a new world filled with entertainment, news and tools made by everyone, that can support social interaction and community development. Social micropayments projects of this type, that let you support bloggers, developers and other creators and enjoy their contributions to the net, are increasing. Click on buttons to reward great web content or add Flattr to your site is one example of this. Customers to Flatr pick how much they like to spend per month on this activity. Then, whenever they see an e.g. Flattr button on any website, that they like, they click it. When they click the Flickr button they get any types of entertainment – gimmick, small play, joke tailor made to the customer. Flickr then count up all of their clicks at the end of each month and distribute monthly spend between everything they clicked on.   Flatr - http://flattr.com/ Big change through small donations

L.O.C.´s famous artists new CD – (http://www.rodekors.dk/kampagner/loc) was sold via the Red Cross website – (www.rodekors.dk) . By bying this the buyer supported Red Cross work. All the profit from the sales was promised to go directly to Red Cros work. The Artist is hereby used as a platform to sponser aid.   Red Cross is dependent of the aid the citizen gives and this is a new initative to the organization. L.O.C. claimed he wanted to help people suffering.

3.7.   Group 7: Energy supply and alternative energy supply (Simpa)

These BM´s have a focus on helping people to energy supply or alternative energy supply. Especially poor people is in focus.

One-line Pitch - Portable solar charging and light through mobile micro-payments for rural African families was introduced to the market by Simpa. Micro payment-control venture formed after 3 years of direct pay-per-use solar energy trials in rural Kenya. 30 combined related-years of mobile-micro payment and rural outreach experience, operating against a proven way to profitably brought affordable power & light to those who could pay with mobile micro-payment.

3.8.   Group 8: Science donation (KB, PCBD)

These BM´s have a focus on donation for science especially in health care or it could also be science in other areas as e.g. culture, where there is a strong need for financing or too little resources for science.

More and more retailers are making social responsibility a key component of their standard business practices. The sale of " Red" and "pink" merchandise to benefit breast cancer research has become familiar to shoppers in a Pavlovian kind of way: Consumers see the now ubiquitous pink products and their brains immediately associate the branding with the Susan G. Komen for the Cure global breast cancer movement.

Likewise, big-box retailers such as Gap, Apple Computers, and Hallmark make their customers see red. Make that (Red), as in (Product) Red, a movement dedicated to eliminating AIDS in Africa. At Gap, half the profits from sales of Gap (Product) Red merchandise go to the Global Fund to help finance AIDS treatment and prevention programs on that continent

The Danish cancer fond runs a lottery with more than 20.000 winners in the lottery pr. year. Hereof 10 winners will have 1 mill. Dkr. It is not possible yet possible to sell the lots online. Therefore those who want to donate and join the lottery must pay by bank transfer or payment at the local post office. The profit from the lottery goes to 3 main areas: Research, Information and Support to patients and relatives

4.   FINDINGS AND DISCUSSION

We now turn to our findings and discussion of persuasive business models.

Table 2 Examples of Persuasive Business models "bundle" of value propositions

Combinations	Businesses
1. Environment and social community focus	British Airways,
2. Business development and social community focus	SBT, Gramham Bank
3. Entertainment and social community focus	Flatr, Red Cross, Kræftens Bekæmpelse,
4. Donation and a physical product	PBCD, Red Cross, TCE, Simpa
5. Donation and a Service	SOS
6. Donation and tax reduction	Red Cross, SOS, CSA
7. None	IMS

4.1. *Combination of different value propositions*

Instead of focusing just on one value proposition e.g. life, business development and gaming we found that that several of the use cases we studied used a kind of combination BM´s strategy to attract and "persuade" the micropayment. They "bundle" different value propositions and BM´s. We found the following value proposition combinations as shown in Table 2.

Generally we saw that there is not used more than two value proposition combination in these BM´s. However in 2 cases it could be argued that there were to some extend up to 3 value propositions given. (PBCD, RAD). Having more than one value proposition combinations seems to be critical to the success of a persuasive BM because the business models can get blurred and fuzzy in the perception of the customer. However the choice of different donation possibilities can be favorable to persuasive business models because this can increase the involvement of the donator.

4.2. *Profit and non profit*

The micro finance BM studied shows a variety of profit and nonprofit BM projects. We found that 3 BMs were pure profit BM projects, 7 nonprofit and 4 BM projects which made a kind of combination of profit and nonprofit BM,s - a multi business model approach (Lindgren 2012).

Table 3 Purpose of persuasive business models related to ecological, social and society behavior

Categories	Businesses
1. Pure Profit	IBS, GB, Flatr,
2. Non Profit	CSA, KB, SBT, TCE, SOS, CCA, HSJD
3. Combination of profit and non profit	BA, PBCD, RAD, Simpa, CBM

5. SUCCESS OF A PERSUASIVE BUSINESS MODEL

We now turn to the question. How can you tell that a BM project combining micropayment ecological, social and/or society capital purpose is or has been a success? We line up some of the dimensions for evaluation, which could be considered as success criteria to such:

Table 4 Success factors of persuasive business models

Success Factors	Achieved	Medium success	Not Achieved
1. Technical success – the technical system in the BM's works and operates with success	IBS, GB, Flatr, CSA, KB, SBT, TCE, SOS, CCA, HSJD BA, PBCD, RAD, Simpa, CBM		
2. User and customer success of those donating the money – they feel they get value for money they donate and/or spend	IBS, GB, Flatr, CSA, KB, SBT, TCE, SOS, CCA, HSJD PBCD, RAD, Simpa, CBM	BA,	
3. Networkpartner success – all or most network partner fulfill their goals and achieve their business and business model intention.	IBS, GB, Flatr, CSA, KB, SBT, TCE, SOS, CCA, HSJD PBCD, RAD, Simpa, CBM	BA	
4. Short term success – the BM project success fulfill a short term need – generating money or change for a specific purpose quickly – life aid e.g.	IBS, GB, Flatr, CSA, KB, SBT, TCE, SOS, CCA, HSJD BA, PBCD, RAD, Simpa, CBM		
5. Long term success – the BM project runs for a long time – and creates sustainable and ongoing business and BM development, social and community BM projects.	Flatr, CSA, SBT, TCE, SOS, CCA, PBCD, Simpa, CBM	IBS, GB, KB, HSJD RAD	BA
6. Relations success – the project and organization builds up a long term relation to those donating the money. The donators keep on donating the BM	SOS, Simpa	IBS, GB, Flatr, CSA, KB, SBT, TCE, CCA, HSJD PBCD, RAD, CBM	BA,
7. Value success – the BM project generates value as profit to the business	IBS, GB, Flatr, SBT, PBCD, RAD, Simpa,		(Non Profit BM's (TCE, CSA, KB, SOS, CCA, HSJD CBM BA (Combined with a Profit based BM)
8. Value success – the BM project generates value as ecological value – ecological success	Simpa, CBM		IBS, GB, Flatr, CSA, KB, SBT, TCE, SOS, CCA, HSJD BA, PBCD, RAD,
9. Value success – the BM project generates value as Social success	IBS, GB, CSA, KB, SBT, TCE, SOS, CCA, HSJD BA, PBCD, Simpa, RAD, Flatr, CBM		
10. Value success – the BM project generates value as Society success	IBS, GB, CSA, KB, SBT, TCE, SOS, CCA, HSJD BA, PBCD, RAD, Simpa, CBM		Flatr,

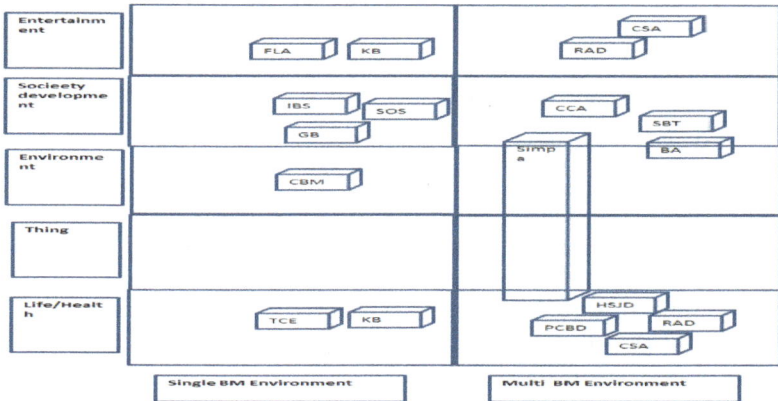

Figure 1 "Persuading tools" related to persuasive Business Models.

As can be seen most of the persuasive business models we studied have been successful in most criteria. – Some have however had minor implementation difficulties in the very early stages. Only very few (BA, GB) and on very few success criteria did not succeed.

It is therefore interesting to have a closer look to how these business models are create, constructed to find out how they achieve their success.

5.1. How do persuasive BM´s get success and how did they change the behavior of people or organizations to donate money?

The different organizations and projects do their outmost to create and attract micro financing. In this work they try to support or change a behavior in favor of micro payment to a project. They try to relate the micro payment to the above mention list of environmental and social capital issues. They do this both as nonprofit and profit based projects. A cross analysis and summary of the BM projects we studied shows some major difference in scope and objectives as seen in Table 4. As can be seen beneath the persuasive BM´s can be related to different "persuasive tools" which they use to "persuade" donators to donate and change behavior.

As can be seen society, life Health and entertainment are the most used "persuading value proposition". In this context we accept that our case material can be biased as we are primarily studying persuasive business models in social, society and ecological business model ecosystems (Lindgren 2012).

Most of the persuasive business models use a multi business model approach (Lindgren 2012), where they use a combination – bundle - between two or more business models to convince the donators to donate money to their business model. The Simpa and KB business even use more than 3 BM´s.

5.2. Microfinance and tangible value proposition

Most persuasive business models seem to come with a kind of **"tangible" value proposition** either **a product**, **a service** or **a process of products and services**. CCA e.g. sells stuff to customers and take the donation out of the profit. RAD use the CD, special coin and signed cd album and Simpa use the solar cell lamp. SOS give the "father" a child which becomes healthcare service, education and the possibility to follow the child and even visit him or her. There is however a big variety in the products and services that is connected to the different micro finance business models. In appendix x we have given an overview of these. Initially there seems to be a strong relations tangible and intangible value propositions inside a BM and between BM´s.

5.3. *Microfinance and intangible value proposition*

Very few micro finance business model use intangible product and services. However BA used this related to saving carbon and supporting community initiatives, but the business model seemed to be a bit fuzzy and unclear both with relation to the saving carbon and to the support fund BM. Further the business model seemed to have difficulties to succeed. It seems as if there was some distance between the donation and the carbon saving together with the funding projects. Therefore it seems as if the donator hereby had some difficulties to relate to the value proposition offered by BA BM.

5.4. *Microfinance and single or multi business model*

Most micro finance BM projects are multi business models, which means that the customer really "buy" a bundle of business models. In BA case e.g. the customers buy a reduction in carbon but also support to a community – and then of course a ticket to a flight travel.

5.5. *Micro finance and technology*

The mobile telephone seems to become more and more important in the setup and operation of BM microfinance projects. Very few of the BM projects were not using the mobile to generate micro financing for their BM´s.

5.6. *Single or Group microfinancing*

Very few of the projects we studied were group funded projects. SBT project was however a group funded project, where the group behind shared the risk of micro financing different social business projects. This means that microfinance project primarily seems to be funded by individuals – "a bottom up approach". However we found some micro finance projects where it was cooperate funded or group funded.

Micro payment and social capital projects is a variety of many individual and few group initiative and investments. Many of the initiatives are carried out by volunteers but others are highly organized and runs as professional businesses. The microfinance business can be focused very short term based aid – focusing at helping people to recover from a catastrophe, hunger – Haiti, Africa. Other initiatives are more long term based – business development, community development, reduction of carbon, reforesting . A lot of these are enabled via micro financing donation in which we give some examples of different type.

5.7. *Micro financing and trust*

We found that persuasive Business models do have to a high degree ti be related to trustfulness. It is therefore vital to the success of the business model that the technology and the organizational structure around the business model are trustful to the donator. If the organizational structure gets too "blurred" the donators will not be attracted. If the relations between more business models gets to complicated then the donator will loose confindence and trust to the business models and not donator to the BM.

5.8. Micro funding and ownership

It was also found that it is important that the donator feels a kind of ownership and involvement to the BM's. This can be established via the BM is strongly matching and related to the values of the donator or the values of the donators business. The persuasive business models must therefore be able to analyze and create a BM which valuepropositions are related to the donators values – otherwise it will be very difficult to achieve success of the business model.

5.9. Micro financing, network and network construction

Persuasive business models are often constructed with a rather complicated network physical network (network of people), digital network (ICT network) and virtual network (network that turn up whenever there is a task or a valueproposition that the persuasive business model have to provide. These networks are often very blurred and not ease to the donator to "see". We found that it is therefore very important to those bidding out the persuasive business models to visualize and communicate very precisely how the network is constructed and which network are involved. It seems to be more important these days as the competition and number of persuasive business models increase.

5.10. Microfunding and nonprofit/profit business model

As seen in model 1 more persuasive business models use the multi business model approach, where they combinate a profit and a non profit business model. Most business we studied in the persuasive business model ecosystem are nonprofit oriented business. However it is often very difficult to see clearly how these so called nonprofit business really are functioning and if the nonprofit organization is just a smart way of doing business.

5.11. Microfunding and user-friendliness

In those cases we studied it was significant that the persuasive business models must be exstremely userfriendly. Especially in the very moment when the donator takes and makes the desion to donate the technology, service and paying process must be very userfriendly, easy and clear.

6. THE OPTIMUM SET-UP IN TERMS OF PERSUASIVE BUSINESS MODELS

Above we addressed some of the issues of ability and hot triggers to persuasive business models and we identify the construction and what characterize a persuasive business model target at changing ecological, social and society behavior.

We now discuss the optimum set-up in terms of these persuasive business models would be.

Our research shows that a persuasive business model at an optimum must adapt a multi business model approach combining different engridience from more than on business model. To illustrate this we use the 7 building block Business model framework (Lindgren 2012) showing two persuasive business model examples.

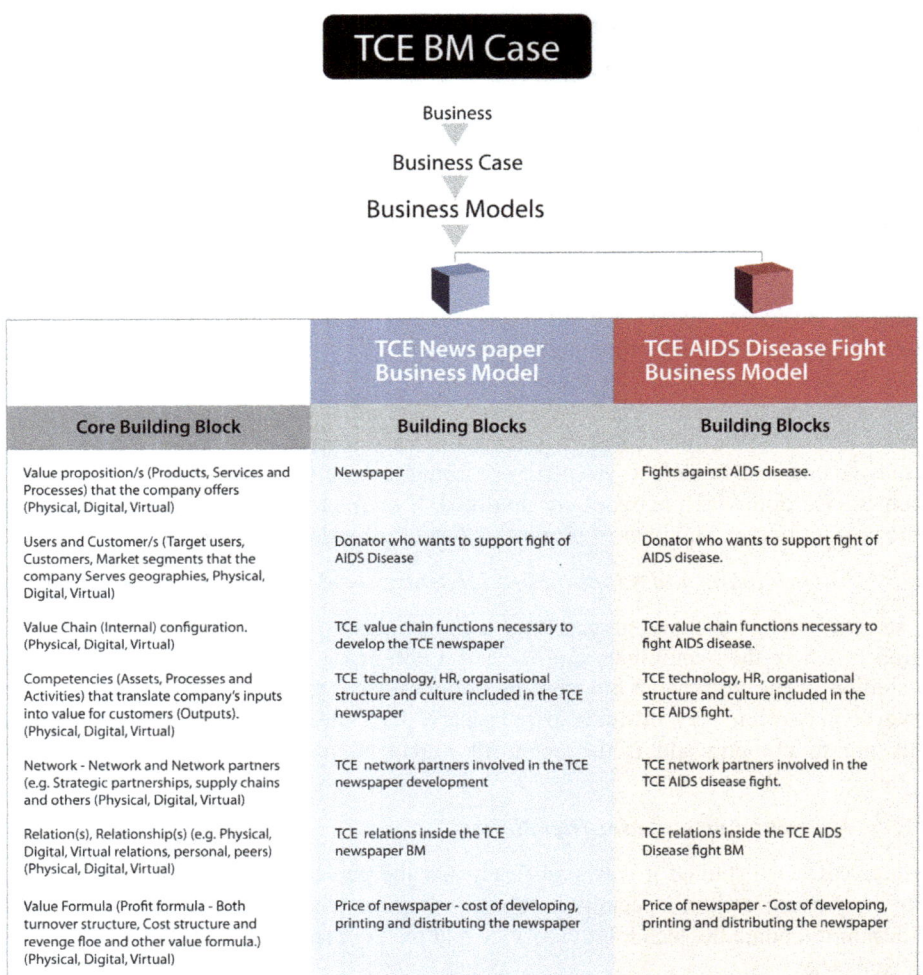

Figure 2 TCE bundle of Business models related to AIDS Disease.

TCE have in this business case chosen to bundle two business models – one with their TCE Newspaper which generate money and is bundle with their second business model – The TCE AIDS Decease fight Business model. The two business models are both offered to the same customer – the donator. In this bundle of business models the donator gets a thing – the news paper – and the promise of TCE fighting the aids disease in one district chosen by TCE. This is how they "persuade" their customers to donate money to their business.

As indicated in the model TCE could have bundle other BM to these two Business Models and TCE have other BM cases in their Business.

KB BM Case

Business → Business Case → Business Models

Core Building Block	KB Lottery Business Model / Building Block	KB Cancer Disease Business Model / Building Block	KB Cancer Information Business Model / Building Block	KB Cancer Patient Support Business Model / Building Block
Value proposition/s (Products, Services and Processes that the company offers) (Physical, Digital, Virtual)	Lottery, possibility to win money, things and services	Research aimed at Fighting Cancer disease	Development of Information material aimed discovering and Fighting Cancer disease at an early stage, preventative	Developing and running Cancer Patient Support
Users and Customer/s (Target users, Customers, Market segments that the company Serves geographies, Physical, Digital, Virtual)	Donator who wants to support research in cancer, support of cancer patients, information about cancer	Donator who wants to support fight of Cancer Disease	Donator who wants to support development of information about Cancer Disease	Donator who wants to support development and running Cancer Patient Support
Value Chain (Internal configuration. (Physical, Digital, Virtual)	KB value chain functions necessary to run the KB lottery	KB value chain functions necessary to handle funding support for cancer research	KB value chain functions necessary to handle information about cancer research, cancer patient support activities, cancer discovery and protection	KB value chain functions necessary to handle Cancer Patient Support
Competencies (Assets, Processes and Activities that translate company's inputs into value for customers (Outputs). (Physical, Digital, Virtual)	KB technology, HR, organisational structure and culture included in the KB Lottery	KB technology, HR, organisational structure and culture included in the KB cancer research funding handling	KB technology, HR, organisational structure and culture included in the KB cancer information activities and handling	KB technology, HR, organisational structure and culture included in the KB Cancer Patient Support activities and handling
Network - Network and Network partners (e.g. Strategic partnerships, supply chains and others (Physical, Digital, Virtual)	KB network partners involved in the KB lottery	KB network partners involved in the cancer research funding BM	KB network partners involved in the cancer information activities	KB network partners involved in the Cancer Patient Support activities
Relation(s), Relationship(s) (e.g. Physical, Digital, Virtual relations, personal, peers) (Physical, Digital, Virtual)	KB relations inside the KB lottery BM	KB relations inside the KB research funding DM	KB relations inside the KB information	KB relations inside the KB Cancer Patient Support BM
Value Formula (Profit formula - Both turnover structure, Cost structure and revenue floe and other value formula.) (Physical, Digital, Virtual)	Price of lottery - cost of developing and running the KB Lottery	Price of newspaper - cost of developing, running and distributing the KB research cancer funding	BM Price of KB information - cost of developing, running and distributing the KB information BM	BM Price of KB information - cost of developing and running KB Cancer Patient Support BM

Figure 3 KB bundle of Business models related to KB Disease lottery BM, cancer research funding, cancer information activity, cancer patient support BM,s .

The KB persuasive business model is much more complicated in its structure than TCE. Firstly it bundles at least 4 business models – the lottery BM, The Cancer funding BM, the Cancer patient support BM and the Cancer Information BM. Although it is said by the KB business that donators like often play the KB lottery knowing that if they do not win then their donated money will be used to something that will be in the could of cancer research, cancer patient support and cancer information then it is much more "blurred" to the donator what his donation specifically go to. KB plays the multi business model approach by bundling more BM´s together and offered this to the donator.

7. Barriers to "jump" to the target behavior and to persuasive business models success as such

How many business models can You play or bundle?. Is there a maximum of business models and a maximum of a bundle of business models. This we have not been able to investigate in our research. However there seems to be a limit on how You bundle business models and how many business model You bundle.

Firstly it seems that persuasive business model bundle with society supporting business models e.g. ecology (Carbon, rainforest protection e.g.), pollution protection, society social issues face harder conditions to success than persuasive business models related to life and health persuasive business models. These persuasive business models could presumably jump barriers to change of behavior or rejection to donation by bundling their business models with other types of business models as entertainment, edutainment, life or health business models.

Secondly it seems that pervasive business models that have smart and easy to operate technology embedded have higher potential for success.

Thirdly it seems that pervasive business models that make the donator personally involved in the donation choice and the donation moment seems to have higher possibilities of success. The personal involvement can either be established via entertainment BM, choice of health care BM (HSJD case – make donator choose between 3 different health care BM´s),

All in all there is several barriers to jump and this is why the BJ Fogg behavior change model can be valuable to change the behavior of people with small steps.

8. "Dark sides" of persuasive business models

Through our analysis of the use cases we also came above some dark sides of persuasive business models

Harnising the social capital in small groups

Persuasive Business models that focus on lending to the poor are e.g. often costly due to high screening, monitoring, and enforcement costs. Group lending advocates that the believe in individuals are able to select creditworthy peers, monitor the use of loan proceeds, and enforce repayment better than an outside lending organization can by harnessing the social capital in small groups. Using data collected from FINCA-Peru (Fiegenberg 2005) exploited the randomness inherent in formation of lending groups to identify the effect of social capital on group lending. She found that having more social capital results in higher repayment and higher savings. She however also found suggestive evidence that in high social capital environments, group members are better able to distinguish between default due to moral hazard and default due to true negative personal shocks.

Leapfrog of traditional banking services and moral

Cell phones have allowed much of the developing world to forgo building an expensive landline infrastructure in rural areas and could now be used to leapfrog traditional banking services. A study by Vodafone suggests that "in a typical developing country, an increase of 10 mobile phones per 100 people will boost GDP growth by 0.6 percentage points." Notwithstanding the potential of mobile banking to expand financial services to the poor, the proliferation of cell phones has had a positive impact on development but also in some cases negative impact. A string of suicides has put micro lending under the spotlight .A messy collision with the realities of local politics made the Indian MFIs think about restriction in microfinancing and micro payment. However MFI finally decided not to and instead focus on talking about something more basic: survival. However politicians from the state of Andhra Pradesh (AP), where microfinance has made the deepest inroads and where SKS has its headquarters – one of the big providers of micro finance- , have held micro lenders responsible for the suicides of 57 people. It was alleged that these people were hounded to their deaths by lenders' coercive recovery practices. MFIs deny wrongdoing. Vikram Akula, SKS's founder, said that although 17 of the 57 women who killed themselves were SKS clients, none was in default so there was "no scope for putting any pressure". Despite this, the state government passed an executive order on October 15th 2005 imposing curbs on MFIs. The order stopped short of capping interest rates, as many had feared, though a subsequent statement by a senior bureaucrat suggested that this remained an option. SKS voluntarily shaved two percentage points off its loan rates in AP, where it had 2.2m borrowers. But it was barely functioning in the state anyway. A series of arrests of so called "field workers" led the business to keep 6,000 staffers idle.

Interest rates too high of persuasive business models

Interest rates of 20-30% may seem high for Micro finance - but so are recovery and loan-servicing costs in remote villages. According to Mary Ellen Iskenderian of the Women's World Banking (Iskenderian 2010), a network of MFIs, a more pressing problem is likely to be over-indebtedness, fuelled by rapid growth in a sector with no formal credit bureaus. This led to that Indian MFIs are now sharing information, pledging not to lend to a person who has already borrowed from three others and to keep total lending to a limit. However smaller lenders have fewer qualms (The Economist 2010)

Persuasive technology to easy to use in Persuasive Business Models

Many persuasive business models have shown to be too easy to use and "too persuasive" leading some donators into serious financial challenges. Above we comment on the easy to lend money via micro loans but also when persuasive business models are related to games and edutainment some donators forget about their economical ability and just "push the button" with out taking their situation into consideration. They are so "persuaded" by the donation situation that the forget about reality.

Persuasive business Models and ethical issues

As more and more persuasive business models move to bundle their business model with edutainment and "life and dead" then they fall into some ethical considerations about the real economical situation and ability of the donator. Still there is no rules to and very few guidelines to these business – how to manage "persuasive business models" in a socieal, society and ethical way.

9. CONCLUSION

Persuasive BM concept is increasingly gaining acceptance within the business world. The concept of persuasive business models is however not new and especially the church have since long been real experts on persuasive business models – to some extent maybe too clever to run these models..

The concept of persuasive business models has evolved within the increasingly globalized world and environment – however no one have yet defined the persuasive business model. Conversely, the evolution of persuasive business models and new persuasive business models is argued to be just on-going and possibly unstoppable in their process. Thus, this paper has been concerned with the relation between persuasive business models and their road to success we found that one of the main roads to their success is their ability to bundle more BM's – the multi business model approach. However when the complexity in the bundle of business models increase the persuasive business models face the risk of "blurredness", which can lead to unsuccessful results.

In the study we also looked for - How the suggested persuasive business models in a business model context might be adapted into other business models and lines of behavior with similar feats? In context of the latter, it is suggested, that there is a potential new business model eco-system on the rise and that other business not using persuasive business models could with preference learn from these and even adapt some of their approach to general business modeling.

In order to counteract these effects and create leverage the idea of performing a persuasive business model with persuasive technology was introduced and the significance of the persuasive business models was related to the huge possibilities. On the other hand, the increased competition and the rapidly developing persuasive business model ecosystem have forced businesses and their BMs to become more agile, dynamic and smart in bundling technics. In this context the persuasive business model is highly related to the evolution of ICT, evolution of network based business models, globalization of BM's and their resources and opportunities availability - hereby involved. However, this raises significant challenges and etical questions in relation to persuasive BM leadership and management.

10. FUTURE EXPECTED RESULTS/CONTRIBUTION

The study has enlightened a first theoretical attempt to persuasive business models. The next step is to initiate a more thorough and empirical based research to clarify the hypothesis, in order to be able to support and test the persuasive business models on a broader scale.

11. REFERENCES

[1] Fiegenberg 2005 FINCA-Peru, Figenberg et all (2010) Building Social Capital Through, Microfinance, Faculty Research Working Paper Series, Benjamin Feigenberg, Department of Economics, MIT, Erica M. Field, Department of Economics, Harvard University, Rohini Pande, Harvard Kennedy School, June 2010, RWP10-019

[2] Fogg, B. J. (2009) A Behavior Model for Persuasive Design ACM ISBN 978-1-60558-376-1/09/04

[3] Fogg, B.J. 2012 Persuasive Technology Stanford University Press, Persuasive Technology: Using Computers to Change What We Think and Do (Interactive Technologies)

[4] Economist 2010, January, Micro financing and social capital

[5] Karnøe, M. et al, "A method for collecting fees in a fee-based document output system" 18784US00 (pat. pending)

[6] Lindgren, P. (2012) Towards a Multi Business Model Innovation Model. / Lindgren, Peter; Jørgensen, Rasmus. Journal of Multi Business Model Innovation and Technology 1 edition River Publisher

[7] Neffics 2012 Delivery D 4.3. European FP 7 IOT program – www.Neffics .eu

[8] Iskenderian Mary Ellen of the Women's World Banking, http://www.forbes.com/sites/evapereira/2011/06/16/the-future-of-microfinance-qa-with-womens-world-banking-ceo-mary-ellen-iskenderian/"The Future Of Microfinance: Q&A With Women's World Banking CEO" www.forbes.com/.../the-future-of-microfinance-q...,
http://webcache.googleusercontent.com/search?q=cache:Zwde5mCGUt4J:www.forbes.com/sites/evapereira/2011/06/16/the-future-16/06/2011

Appendix 1. Persuasive Business Model Use cases

1. **British Airways' - Unique One Destination Carbon Fund (BA)**

    Scope: Environment and social responsibility – non profit, CSR and branding oriented related to BA

    Objective: Reduce Carbon, Environmental protection, Social wellfare, social responsibility, community development, nonprofit, branding

British Airways' decided in 2008 to establish a "unique One Destination Carbon Fund" where they toke customers flight ticket donations and used them to support energy efficiency and renewable energy projects in communities in UK. BA was targeting parts of UK that BA thought needed the most help to improve their economic and social well-being. The funds were managed a not-for-profit charity fund, using the UK Carbon Reporting Framework as quality insurance. The donation funded a range of projects from helping install solar hot water in community swimming pools, small scale wind turbines for schools or energy efficiency measures in social housing. BA ensured their customers donation money went to make projects that gave real difference in communities across the UK – via connecting each project to support saving carbon and so its was good for the environment as well as local communities too. BA hereby related carbon saving with social projects in local communities.

Comment: Has not yet in particular been very successful in amount of donation, but to some extend on CO2 and social impact.

2. **Indian, Bangladesh and South African based Micro finance (IBS)**

    Scope: Mobile Micro financing business model to poor people at the bottom of the pyramid – profit oriented

    Objective: Business development, development of life, profit, new markets

Access to capital and financial services is a problem with in developing countries. Lack of access to banking services hinders economic development and gives the poor no option other than the informal cash economy, leaving them vulnerable to risks and without any means to efficiently save or borrow money. Higher saving rates have proven to make more capital available for investment in development.

> "What we're finding from the evidence from economists is that actually greater access to financial services improves economic growth,"
>
> Jeremy Leach of FinMark Trust, an NGO that promotes financial services for the poor.
>
> "For many poor South Africans, the system offers a first step into a world that can help them save, send, and receive money. With a few key

punches, they can send money to a relative or pay for goods without ever seeing a paper bill—a benefit in a country with a high crime rate,"

Source "Nicole Itano of the Christian Science Monitor"

Consultative Group to Assist the Poor (CGPA) estimated in 2008 that 80 percent of people in least developed countries are unbanked. The term unbanked refers to people who do not use simple banking services that the developed world takes for granted, such as checking and savings. Barriers to conventional methods of banking include lack of education, illiteracy, high fees, and proximity to banking facilities.

(Source.: GPFI http://gpfi.org/about-gpfi/partners/consultative-group-assist-poor)

Cell phone banking have revolutionizes financial services for the poor. A woman manages e.g. a village from one cell phone for a project in rural Bangladesh. Another woman manage equally in rural area in South Africa food supply and banking. SKS, Indias biggest microfinance institution (MFI) expects that an infusion of private capital will spur even greater growth in credit to India's rural poor, where nearly estimated 27m of whom are already microfinance clients (MFI 2010). Banks and cell phone companies are taking advantage this expansion of cell phone use in developing economies to extend financial services to roughly 2 billion people, who use cell phones but lack bank accounts. The "dark sides" are comment later.

Comment: Particularly successful in profit term and impact but with some "dark sides"

3. The Social Business Trust (SBT)

Scope: Micro financing business model to develop social businesses – profit oriented

Objective: Growing and developing social businesses, movement of philanthropy and socially responsibility investment

Social Business Trust ("SBT") is a partnership of six world class businesses, who have come together to combine their resources and expertise to help accelerate the growth of social enterprises. SBT believes there are a number of social enterprises capable of scaling up their operations on a regional and national level and has a clear and ambitious goal: to help transform the impact of social enterprises and thereby improve the lives of up to a million people in the UK. SBT's strengths and capability come from a combination of the partners' commercial and industrial experience; their insights into the needs and sensitivities of social enterprises; and their operational expertise and access to growth capital.

SBT give out "small micro grants" and pro-bono support, with the aim of rapidly growing more than 20 social businesses. The management team behind Bain & Company had initially worked together, supporting social business to scale up. – (http://www.socialbusinesstrust.org/about-us)

Comment: Very succesfull in profit term and impact

Since inception, SBT has made six investments in UK based social business:

Women Like Us [1], The Challenge Network [2], Moneyline [3], London Early Years Foundation [4] Inspiring Futures Foundation [5], Bikeworks

Social Business Trust is part of a wider movement of venture philanthropy and socially responsible investing. SBT was highlighted as a best case study in the 2011 Giving White Paper as an example of how to donate professional and specialist skills.

4. Grameen Bank (GB)

Scope: Mobile micro financing business model - Profit oriented

objective: Fight against poverty, homelessness, destitution and inequality

Microfinance made headlines when Grameen Bank – one of the first microfinance bank - founder Muhammad Yunus won the 2006 Nobel Peace Prize. The cost of the small transactions involved in microfinance— savings accounts, money transfers, and loans to the poor—have been an obstacle. The use of cell phones has verified to cut the cost of such transactions, making widespread microfinance more efficient. A CGAP study funded by the Bill and Melinda Gates Foundation, found that cell phone banking was potentially six times cheaper for routine banking transactions.

"Grameen Bank is a message of hope, a programme for putting homelessness and destitution in a museum so that one day our children will visit it and ask how we could have allowed such a terrible thing to go on for so long", he said. Source.: http://www.grameencreativelab.com/events/worldwide-social-business-day-2012.html

Comment: Very successful both in profit term and in impact.

Muhammad Yunus, winner of the 2006 Nobel Peace Prize and Managing Director of the Grameen Bank since 1985, has been widely recognized as the originator of the use of microcredit as a powerful tool in the fight against poverty and inequality. The Bangladesh-born Fulbright Fellow has a vision of the global eradication of poverty.

5. Flatr (FLA)

Scope: Entertainment micro financing business model

Objective: Entertainment, social capital

The latest years there has grown a new world filled with entertainment, news and tools made by everyone, that can support social interaction and community development. Social micropayments that let you support bloggers, developers and other creators. Click on buttons to reward great web content or add Flattr to your site. Customers to Flatr pick how much they like to spend per month on this activity. Then, whenever they see a e.g. Flattr button on any website, that they like, they click it. When they click the Flickr button they get any types of entertainment – gimmick, small play, joke tailor made to the customer. Flickr then count up all of their clicks at the end of each month and distribute monthly spend between everything they clicked on.   Flatr - http://flattr.com/ Big change through small donations

Comment: Successful to some extend on profit and impact.

6. Simpa Networks (Simpa)

Scope: Energy saving business model to poor people – profit oriented

Objective: Energy access to Poor people, safer energy, cheaper energy, healthier energy

Simpa Networks was founded by Paul Needham, InfoTech entrepreneur with two successful exits and a dozen years senior leadership experience, most recently at Microsoft Corp. Jacob Winiecki, recognized thought-leader on energy access and former operational specialist with Arc Finance, with over 5 years experience in microfinance and energy access across India, Sub-Saharan Africa and East Asia, and Michael MacHarg, MBA, former micropayments advisor to Arc Finance and former Acumen Fund consultant with over a decade of experience leading global social enterprises across the spheres of energy, finance and health joined the Simpa Networks. Simpa Networks saw that worldwide, approximately 1.6 billion people have no access to electricity and another 1 billion have extremely unreliable access. Without ready access to electricity, the poor depend on kerosene lanterns and battery-powered flashlights for light. Kerosene lanterns are dangerous, dirty, and dim. Worse, they are very expensive to operate. And yet, in most markets, kerosene lanterns are the preferred lighting system. For a person with little or no savings, no access to formal credit, and low and uncertain income, the selection of kerosene lighting is eminently rational. Kerosene lighting has a low initial purchase price and offers a flexible pattern of expenditures over time. The consumer can choose how often and how bright to burn the lantern, and often chooses to forego light entirely for periods when income is unavailable. The kerosene light – with its high operating costs, its many dangers to health and home, its poor quality light and noxious fumes – has been the best choice available. The poor people can't break the cycle of poverty because they can't take advantage of the myriad productive uses of energy. Access to energy is essential for a family's economic livelihood, health, safety, educational achievement, and quality of life – "It's Expensive to be Poor".

Many individual consumers in many emerging markets are making less than $10/day, with the poorest spending up to 30% of their income on inefficient and expensive means of providing light and accessing electricity. Worldwide, low income consumers spend about $38 Billion per year on kerosene for light, another $10 Billion on cell phone charging. Simpa estimates that there is likely a $100B global opportunity for small scale distributed energy solutions, with no clear market leader.

Modern energy systems that meet these lighting and basic electrification needs are on the market for $200-$400 retail. These systems typically include a solar panel, battery, charge controller, at least 3-4 lighting points, a mobile phone charging port and power for charging or powering small DC devices. These solar home systems have proven to be very desirable to consumers who immediately recognize the health, educational, and income generating benefits. Yet households cannot afford the high upfront cost of a quality solar energy system and thus remain locked into expensive fuel-based lighting and battery charging fees. Over the 10 year useful life of a quality SHS, households will spend $1500-$2000 on kerosene, candles, batteries and phone charging. They are paying more than they need to, because they are poor and because their incomes are low and unpredictable. Underscore this fact: In our launch market, India, as in most developing country markets, the low income consumer can actually afford a small solar home system if only they could pay for such a system over time, in small, irregular, and user-defined increments. That is, if the pricing model matched the pricing model they are already using for kerosene, candles, batteries, and phone charging.

Comment: Very successful both on profit and impact.

One-line Pitch - Portable solar charging and light through mobile micro-payments for rural African families was introduced to the market. Micro payment-control venture formed after 3 years of direct pay-per-use solar energy trials in rural Kenya and 30 combined related-years of mobile-micro payment and rural outreach experience, operating against a proven way to profitably bring affordable power & light to those who can only pay with cash-on-hand.

7. The charity button machine (CBM)

Scope: Donation Business Model – non profit

Objective: Saving nature, Environment protection, Social responsibility, nonprofit, branding

By pushing on the charity button at the bottle automat (Appendix x) at the largest retail chain in Denmark Coop (www.coop.dk) it is possible to donate bottle recycling money to wellbeing projects in WWF – (www.wwf.dk) and Red Barnet – (www.savethechild.dk). About 1100 bottle automats have this extra button in Denmark. It is possible to donate the whole or part of the recycling bottle money. This is done by pushing the Wellbeing button first, and when the customer reaches the amount they want to donate then they push the charity button. Afterwards the rest of the bottles and cans are delivered and the pay button is pushed and a cash ticket can then be changed to cash at the cash register.

Comment: Very successful related to donation.
   The system was tested in 2007 and from September 2008 it has been up and running in all Coop´s retail shops. The donation to Red Barnet and WWF Verdensnaturfonden was in 2010: 7.980.644 Dkr. And in 2011: 7.722.483 DkR.

8. The TCE project (TCE)

Scope: Donation Business Model - nonprofit

Objective: Health care, Social responsibility, community development, nonprofit

TCE was developed as a program by Humana People to People. TCE means Total control of the Epidemic (HIV/AIDS-epidemy). The objective of TCE is to reach total control with AIDS in Southern Africa. TCE builds upon the belief that people by them self can win the battle of AIDS and prevent the HIV infection, while all other people can participate in the battle via small donation. The battle against HIV/AIDS is carried out very systematically. A country is divided up into districts and then subareas. One subarea includes 100.000 people. In every subarea TCE employs 50 locale TCE Field Officers within 3 years. Their task is to free the area for more AIDS spreading. TCE Field Officers walks form person to person and inform about HIV/AIDS, and how people can be tested and protect themselves. The Field Officer develops healthcare programs for ill people, pregnant women and children without parents. Further the Field Officer establish nabour support groups, working groups and collaboration with health care organizations together with personal advice to families.

The scope is to fight HIV and AIDS with the "TCE army", who do not work with weapons, but knowledge, innovation, persistence and sticking together. Money is collected for the program more times every year via small donation. This work is especially done by school pupils and teachers, and by selling the TCE-newspaper.

Comment: Successful to some extend in collecting money for the project. Successful related to the impact.

9. "The Churchs Cross Army" (CCA)

Scope: Donation Business Model - nonprofit

Objective: Social responsibility, community development, religious charity project, branding

The Church Cross Army collects stuff that people do not any longer want to keep and then sell these items in special shops. The money collect is then donated for certain social and community projects or send to rural areas for help and support. The products are sold by voluntiers and then small money donation from each product is then used for donation. The system has a kind of double donation as people come with their stuff and give it to CCA. Then people can come and buy the stuff for a low price. The system also have a kind of recycling impact on society as old stuff is reused.

Comment: Successful on collecting money and impact.

10. Cancer reduction "Kræftens bekæmpelse".: (KB)

Scope: Donation via gaming Business Model – non profit

Objective: Health care – specific cancer, Social responsibility, nonprofit, ...

The Danish cancer fond runs a lottery with more than 20.000 winners in the lottery. Hereof 10 winners will have 1 mill. Dkr. It is not possible yet possible to sell the lots online. Therefore those who want to donate and join the lottery must pay by bank transfer or payment at the local post office. The profit from the lottery goes to 3 main areas: Research, Information and Support to patients and relatives

Comment: Very successful and high impact

The big amount of money that the winners get makes a lot of people dreaming and interest to play/donate a large amount of money. The players get a feelling of a 'win-win situation'. Although they don't win, they are always sure, that cancer patients and their relatives will get advantage out of the lottery. In 2011 the lottery gave a total profit of 58,5 million Dkr.

94   P. Lindgren et al.

11. Breast cancer aid –   "Pink breast cancer donation (PBCD)

Scope: Donation via product or "at the cash register" donation Business Model – non profit

Objective: Health care – specific cancer, Social responsibility, nonprofit, branding

More and more retailers are making social responsibility a key component of their standard business practices. The sale of " Red" and "pink" merchandise to benefit breast cancer research has become familiar to shoppers in a Pavlovian kind of way: Consumers see the now ubiquitous pink products and their brains immediately associate the branding with the Susan G. Komen for the Cure global breast cancer movement.

Likewise, big-box retailers such as Gap, Apple Computers, and Hallmark make their customers see red. Make that (Red), as in (Product) Red, a movement dedicated to eliminating AIDS in Africa. At Gap, half the profits from sales of Gap (Product) Red merchandise go to the Global Fund to help finance AIDS treatment and prevention programs on that continent.

Unlike pink breast cancer merchandise, (Red) products are not necessarily red. But they are hip, such as the popular girl's white angora- blend hoodie for sale at Gap. The word "ado(red)" is written in pink across the front, with a heart in place of the letter "o."

Pink week where another initiative were the retailers and their employees support the donation initiative. At the cash register before paying for the goods the customer is asked if they would donate some money for the breast cancer initiative.

Embracing Better Business Practices - Gap cites, as its reason for participating in the program, that companies in today's world should go beyond the basics of ethical business practices and embrace responsibility to people and to the planet.

It's just the right thing to do, according to Senior Vice President of Social Responsibility Dan Henkle, "it also unlocks new ways for us to do business better."

Comment: Successful on collecting money and impact.

12. Red Cross.: "Artist donation"   (RAD)

Scope: Donation Business Model via an artist cd or concert tickets

Objective: Health care, Life aid, ... nonprofit, profit, branding, ...

L.O.C.'s famous artists new CD – (http://www.rodekors.dk/kampagner/loc) was sold via the Red Cross website – (www.rodekors.dk) . By bying this the buyer supported Red Cross work. All the profit from the sales were promised to go directly to Red Cros work. The Artist is hereby used as a platform to sponser aid.   Red Cross is dependent of the aid the citizen gives and this is a new initative to the organization. L.O.C. claimed he wanted to help people suffering. By buying the album in a digital version each donation will give 30Dkr. To Red Cross. If one buys both album 1 and 2 in a dobbelt album they donate 100 Dkr. and if they buy for 500 Dkr. they get a unik Danish Natial Bank developed Red Cross/L.O.C: coint and a singnated dobbelt albumn. The coin is only produced in 500 pieces and the physical dobbeltalbum is also only published in a limited amount.   The cooperation with L.O.C. runs for one month and can be bought via L.O.C.s and Red Cross own webplatforms, in CD retail shops, both physical and digital streaming-service platforms.

13. Concerts – Save Africa. (CSA)

Scope: Donation Business Model via group of artist and concert event + cd

Objective: Health care, Life aid, …nonprofit, branding, …

Give a hand to Africa was the title of a popsong from 1985 written by Nanna and made in a studio together with 1980'ties most popular Danish pop- and rock musicians as an aid cd for hungry people in Africa . The inspiration came from Bob Geldorf and Midge Ure from 1984 and the American version We Are the World written by Michael Jackson and Lionel Richie. Nanna won the competition to wright the song as red cross had made a completion amongst pop- and rock song writers, who should have the opportunity to write the song. The song was played at television-transmitted support concerts Rock for Afrika.

Comment: Highly successful on collecting money and impact

14. HSJD – The Spanish Hospital

Scope: Donation Business Model via product sells

Objective: Health care, Life aid, nonprofit, branding, …

The hospital Hospital Saint Juan de Dieu (HSJD) belongs to the Hospitaller Order of Saint John of God and is a private, non-profit hospital. The order is represented in more than 50 countries and has almost 300 healthcare centers worldwide. HSJD is located in Barcelona, Spain, and is a children and maternity care center. HSJD is a university hospital connected to the University of Barcelona and they are also associated with the Hospital Clinic of Barcelona, which helps the hospital to provide top-level technological and human care. HSJD is 95 % financed by the Catalonian public system and the remaining 5 % comes from private investments. One of HSJD challenges relates to finance, because of the increased cutback in the revenue from the Catalonian public system, since the Spanish economic is under a significant pressure. In regard of that one of HSJD goal is to get a stronger foundation of finance and also be more efficient, which HSJD wants to enable though innovative solution and participate in innovative partnerships and networks – herunder micro payment.

In HSJD Hospital Saint Juan de Dieu – (www.hsjdbcn.org) it is possible to buy at the hospital shop several products with the logo of the hospital. When the customer is asked to pay they are also asked to decide whether the profit of the product should go to one of 3 healthcare project.

1. Help for Children from Somalia to be operated at the hospital in Barcelona Spain,
2. Help to children suffering a certain decease
3. Help to poor children in spain to go to dentist.

Comment: Successful on collecting money and impact

15. SOS - Children fond

Scope: Donation Business Model

Objective: Social responsibility, health care, new hope, education, nonprofit,

Many children that have lost their parents through war, illness, hunger e.g. lives alone in deep poverty without any possibility to get education, food or some chariness. As a socalled SOS-father donators can help a child for only 7 Dkr. Pr. day. Hereby the child will get a new home, family, healthcare and education. The fond promise that only 10 % will be used for administration and no one will be tied up to the agreement forever. This means that the SOS-farther can stop payment at any time. The full amount of money can be used as tax reduction. Further it is possible to follow at a website what the money is used for. Further the SOS-Father will receive a SOS-news magazine.

A. Product, service and process related to micro finance business models.

Project	Product Physical	Product Digital	Product Virtual	Service Physical	Service Digital	Service Virtual	Process of product and service
1.BA (BA)	Ticket	Ticket	CO_2 reduction	Community investment			
2.IBS (IBS)	Money	Bank account			Bank service		Money and bank service
3. The Social Business Trust (SBT)	Venture Money for micro Social capital project investment			Board membership and professional help to set up business			Support during the process of setting up and running the business
4.Grameen Bank (GB)	Money	Bank account			Bank service		Money and Bank service
5. Flatr (FLA)	Money to social capital responsible business, products and projects	Flatr logo on the vendors website to click on Track of clicks and accounting			Entertainment Track of clicks and accounting		
6. Simpa Networks (Simpa)	Solar cell lamp and source operated by mobile telephone	Mobile lightning Track of use and accounting		Lightning	Track of clicks and accounting		Mobile service and light
7. The charity button machine (CBM)	Get rid of bottles and cans Money back from bottle and cans	Money will be sent to the WWF and Red Cross					

Persuasive Business Models 97

8. The TCE project (TCE)	Newspaper Money send and used by The TCE project			TCE operation – information, workshops e.g. in the TCE areas			TCE process of "fighting the AIDS"
9. "The Churchs cross army" (CCA)	Get rid of old and unwanted stuff			Handling and taking care of the old "stuff" and sending the "profit" to the projects			
10. "Kræftens bekæmpelse" (KB)	Lottery lot If winning Money Promise that the money will be spend according to the aim of the lottery and KB´s statements			Funding service to researchers, support groups and information service activities			
11. "Pink breast cancer donation" (PBCD)	Products if RED action Possibility to donate via the cash register if Pink action. Security that the money will be donated according to the objectives and purpose	Digital donation and Payment service		Taking care of the money handling so that the money falls in to the right hands according to the purpose of the "red" and "the Pink" project			
12. Red Cross.: "Artist donation" (RAD)	CD, Special coin, Special signed Album	Digital ordering and Payment service		Having LC to sign the albums Transfer the money to the right purpose and organization			
13. Concerts – Save Africa. (CSA)	Concert, Concert tickets	Digital ordering and Payment service		Transfer the money to the right purpose and organization			
14. HSJD – The Spanish Hospital	A product in form of a HSJD moscot, a book			Taking care of the money handling so			

(HSJD)	for children, and other products from the HSJD shop			that the money falls in to the right projects of the 3 projects			
15. SOS - Children fond (SOS)	A certificate that the donator is a "Father"	Digital ordering and Payment service Track and trace of the money flow		Taking care of the money handling so that the money is given to the child sponsored			Track and trace of the how it is going with the child sponsored

BIOGRAPHIES

Peter Lindgren is Associate Professor of Innovation and New Business Development at the Center for Industrial Production, Aalborg University, Denmark. He holds B.Sc in Business Administration, M.Sc in Foreign Trade and a Ph.D in Network-based High Speed Innovation. He has (co-)authored numerous articles and several books on subjects such as product development in network, electronic product development, new global business development, innovation management and leadership, and high speed innovation. His current research interest is in new global business models, i.e. the typology and generic types of business models and how to innovate them.

Morten Karnøe is Professor in humanistic knowledge processes and culture-driven innovation at the Department of Learning and Philosophy at Aalborg University, where he also is Director of the Interregional Centre for Knowledge and Educational Studies. His research is primarily focused on technology, innovation and culture. In recent years his major area of research has been culture-driven innovation, facilitating mergers between cultural knowledge and specific innovative areas. He has published widely, both nationally and internationally, on a number of areas such as innovation, technology diffusion and economic development. Furthermore Morten has extensive experience in project management and guidance at both national and international level. From 2008 (initiated 2009) to 2010, Morten build, secured funding for, and took leadership on an EU Interreg IVA project – IKON - incorporating 48 institutional partners, and more than 300 individuals, in Norway, Denmark and Sweden (see: http://www.ikon-eu.org/). Management tasks involved all aspects from accounting over HR to research leadership. Morten has in recent years built partnerships with several international companies. He has been visiting professor at Stanford University in 2011 1nd 2012, and was also in 2011 appointed Aalborg University expert in innovation

Mark Nelson is a senior researcher at the Persuasive Technology and Peace Innovation Labs at Stanford University. Mark Nelson is project lead at EPIC Global Challenge, researcher and practitioner at Stanford University's Persuasive Technology Lab, and founding member of Stanford's Peace Innovation Lab. In addition, Nelson is founder of

Peace Markets and advisor at Gumball Capital. He was previously advisor at Vipani.org and Panango.org and an independent business consultant.

B. J. Fogg is a Professor at Stanford University. Dr. BJ Fogg directs the Persuasive Tech Lab at Stanford University. A psychologist and innovator, he devotes half of his time to industry projects. His work empowers people to think clearly about the psychology of persuasion and then to convert those insights into real-world outcomes. BJ has created a new model of human behavior change, which guides research and design. Drawing on these principles, his students created Facebook Apps that motivated over 16 million user installations in 10 weeks. He is the author of Persuasive Technology: Using Computers to Change What We Think and Do, a book that explains how computers can motivate and influence people. BJ is also the co-editor of Mobile Persuasion, as well as Texting 4 Health. His upcoming book is entitled The Psychology of Facebook. Fortune Magazine selected BJ Fogg as one of the "10 New Gurus You Should Know".

Global Business Model – a step into a liquid business model

Kristian Bonde Pedersen[1], Kristoffer Rose Svarre[1], Dmitrij Slepniov[1], and Peter Lindgren[2]

[1]*Center for Industrial Production, , Aalborg University, Denmark.*
[2]*Department of Mechanical and Manufacturing Engineering, Denmark.*

Received: 12 February 2013; Accepted: 25 February 2013

Abstract

The purpose of this paper is to uncover how globalization affects the concept of business model (BM) and business model innovation (BMI). BM and BMI theories will be related to two business cases—IBM and Boeing—to support the authors' conceptualization of the future evolution of BMs and BMIs in relation to globalization.

The paper proposes that BMs should not present a snapshot of a business's present or future way of operating BMs when compared to globalization. It is deemed that a definition of a global business model (GBM) is needed to embrace the opportunities and challenges created by globalization. It is furthermore suggested that a GBM should be agile and dynamic in order to create a sustainable GBM which will suffice in the new globalized world. A liquidized BM that refers to the non-static appearance of future BMs is proposed.

The amount of literature concerning BMs and globalization has increased in recent years. However, literature based on the relation among the two areas is considered limited. The paper attempts to fill this gap in literature.

The paper focuses on the following research question:

- "How can a global business model be defined?"

Keywords: Global business model, Liquid business model, Globalization, Business model, Business model innovation, Sustainable business models

1. Introduction—Why business models innovate?

Today, the term 'globalization' is everyday speech and the increased competiveness of emerging economies intensifies the competition on businesses in the global market and, in particular, their BMs. Businesses acting in the global market are increasingly exposed to challenges from other businesses producing similar products and services with comparable quality, but at lower costs, and by using similar BMs. Accordingly, the focus on BMs as a competition parameter has increased drastically in recent years (Taran Y.E., 2009), (Fielt, 2011). The emphasis on the BM concept has been in focusing on adjusting the BMs to comprehend and oblige the complexity and increased competitiveness of the new global playground, which is moving towards a more individualistic (Friedman, 2007) business network and the evolvement of multi BM businesses (Lindgren 2012). Any business consists of one or more business models. This raises questions to, how businesses can distinguish themselves by innovating their individual BMs in the highly competitive global market, while still staying on the competitive edge. This question also implies increasing importance of business model innovation (BMI), which is related to the incremental and radical changes and reinvention of BMs in order to stay competitive. (Peng, 2010)(Lindgren, 2011)

Focusing only on internal BMI is no longer deemed sufficient, and the focus is thus changing towards a more holistic BMI approach taking in the business network as domain for the BMI. This is known as an Open Business model innovation (OBMI) (Chesbrough, 2007) (Daft, 2010) (Lindgren, 2011). In the ever-changing and increasingly competitive global market, which according to Friedman (2007) is a result of the on-going process of globalization, Chesbrough (2007) emphasizes the need for more OBMIs, including developing other businesses models. However, as ICT is developing and accelerating globalization, it is also allowing for new ways of collaboration (Vervest 2005, Shenkar 2006), but it is questionable if the existing terminology of OBMI is sufficient for the futuristic GBM landscape.

The focus of the present study is to bridge the consequences of globalization and the BM concept to define a GBM. This is done in order to emphasize the opportunities of globalization and thereby guide businesses towards the mindset of a liquid BM and BMI.

2. DESIGN/METHODOLOGY/APPROACH

The methodology applied in this paper is structured around deductive reasoning. First, a theoretical background is presented to provide a foundation for understanding the BM concept and the forces driving the globalization. To exemplify globalization's impact on today's businesses and their BMs, two business cases are included. To stress the general importance of the issue(s) presented in this paper, the cases represent two different businesses. Both cases are chosen based on their relevance for the emerging globalization as they are both contemporary. The information for the cases is gathered through desk research. Based on the cases and the theoretical review, a hypothetical definition of the GBM is formulated. An actual test and confirmation of the hypothesis has not yet been conducted.

Accordingly, this paper will, from a theoretical perspective, firstly define globalization, highlight consequences of the on-going globalization process, and consider how this plays a significant role for businesses operating in the global market. Secondly, the BM concept will be defined based on existing literature. Furthermore, BMI will be related to the existing BM literature. In order to get inputs from the real business world, two business cases, respectively IBM and Boeing, are presented. Finally, the theoretical review and the two business cases will be the foundation for discussing and relating the BM concept and globalization, in order to identify the next step towards defining a GBM.

3. WHY DO BUSINESSES NEED TO INNOVATE THEIR BUSINESS MODELS?

In the following section, the theory on firstly globalization and secondly the BM concept will be reviewed in order to clarify their definitions.

3.1. *GLOBALIZATION*

Globalization can be defined as the free movement of goods, services, labor and capital (Wolf, 2004). Friedman (2007) argues that globalization is an on-going process. He compares it to a monster truck with the gas pedal stocked (Friedman, 2007). It can be discussed whether the globalization process has been on-going since the start of human history, as human beings have carried out businesses across borders for centuries (Peng, 2010). However, the globalization process has been accelerated with the emergence of ICT (Friedman, 2007). Technological development has diminished cost of transportation and communication, which are driving globalization towards an integration of economies (Wolf, 2004). but it has also enabled "things" to act as contributors and actors in the global business environment along with humans (Lindgren 2012).

Friedman (2005), states that "the world is flat". This provoking statement might be interpreted as a way of saying that the global market is homogeneous. Levitt (1983) explains the homogeneous global market as an underlying force which drives consumers to want the same products and services, if they are made available at competitive low prices and at equal quality. Friedman (2005) is, however, more likely referring to the increasing accessibility of the global market. The world today is becoming flat in terms of market accessibility, but it is not flat in regards to diversified needs and cultural, economic and social distances (Ghemawat, 2001). The perception that the global market is becoming a homogeneous mass might be mistaken by businesses' increasing capabilities to deal with customers' differences. Globalization has unarguably opened market opportunities significantly but the competition has concurrently been increased (Friedman, 2007) (Daft, 2010). While businesses around the globe are competing for the same customers, they are also competing for the same resources.

As markets are increasingly globalized, the accessibility of resources and technology has also increased. The technological development and the decreased transportation costs have opened opportunities for utilizing resources and network partners from all around the globe. Due to the open market situation and the increased competition, businesses are forced to seek to optimize their BMs. Not only internal processes are the focus of optimization, but businesses and BMs are also optimized via outsourcing and offshoring , to e.g. low cost labor areas, which enables them to take advantage of advanced technological capabilities from network partners' BMs from anywhere – anytime (Lindgren, 2011).

Until the millennium, globalization was led by governments and multinational corporations (Friedman2, 2005). Today, the global business is accessible for everyone and even small and medium sized businesses as well as individuals purchase and sell across national borders. This interconnection, created by ICT, among individuals, businesses and nations all around the globe is, in this paper, referred to as the global network.

The new environment demands new competitive parameters, where the focus on internal optimization is no longer sufficient (Daft, 2010). The focus has thus changed from businesses competing against businesses on BMs, to networks competing against networks on BMs (Lambert & Cooper, 2000) (Daft, 2010)(Lindgren 2012). With the new network perspective, new forces have emerged. Companies are not only benefiting from drawing on the networks' capabilities, resources and BMs in general, but they are also interdependent on the other network partners' BMs. This means that businesses are no longer autonomous entities that can focus only on the internal aspects of the business but are forced to comprehend and contribute to the network. (Sirkin, 2008).

To comprehend this new business environment, it is suggested that companies do not only base their BMIs on the internal aspects of the network, but rather focus on the BMIs in collaboration with the entire global network, where not only businesses are potential collaborators, but even single individuals can be potential contractors. It is thus relevant to clarify the development and the evolution of the BM concept in regards to globalization. Accordingly, it is relevant to firstly sum up and define what globalization is, before approaching how globalization affects the BM concept.

The increased complexity of the global market allows businesses to act across borders and utilize resources from around the world, while being challenged by increased competitiveness. This has changed the global business landscape from consisting of businesses to include individuals as businesses and BMs, existing in global business and BM based on networks. Thus, globalization can be defined based on Sirkin (2008) and Lindgren (2011): the competition with everyone from everywhere for everything (Sirkin, 2008) – at any time (Lindgren 2011).

3.2. *BUSINESS MODELS*

The first discussion on BMs can be traced back to an academic article in 1957 (Fielt, 2011). However, the concept did not gain acceptance until the mid-1990's (Fielt, 2011). The question—what is a BM? —has been raised, discussed and answered by many researchers in the last decade (Fielt, 2011). However, the answer is inconclusive. Porter (2001) argues that a "definition of a BM is murky at best. Most often, it seems to refer to a loose conception of how a company does business and generates revenue. ...." p. 73 (Porter 2001). Morris *et al.* (2003) have, after reviewing existing theory on business models, in the period of late-1990's to 2003, concluded that a company's potential creation of value cannot be explained from the BM model theory, and that "a general accepted definition has not yet emerged" p. 8 (Fielt, 2011). However, Osterwalder *et al.* (2004) have summed up academic work on BMs from the past 20 years, and stated that a definition of a BM broadly related to a blueprint of how a company does business (Osterwalder, Pigneur, & Tucci, 2005). They further argue that a BM is a set of elements, which will be referred to as building blocks, that by their interrelation expresses the logic of how a company earns money (Osterwalder, Pigneur, & Tucci, 2005).

Osterwalder & Pigneur (2010) have, in the past, been recognized for their approach to the BM concept (Fielt, 2011). Thus, the BM concept is, in this paper, inspired by Osterwalder (2010) but developed further, referring to the following definition: A BM describes the rationale of how a business creates, captures, delivers and receives value.

Important to notice is the distinction between business (Abell 1980) and BMs, as a business is considered to have one or more BMs i.e. the multi business model approach (Lindgren 2012) (Lindgren 2012). Furthermore, all BMs can be referred to "As-Is" BM—already operating in the market and "To-Be" BM—being innovated to the market (Lindgren 2012).

From the BM concept's infancy until today, it can be documented that the BM concept has naturally evolved and changed in relation to the BM context. Globalization has increased businesses' interdependency and today businesses are connected in networks (Daft, 2010) (Peng, 2010). Thereby, it is possible to utilize resources across businesses and its BMs' boundaries in order to strengthen the competitiveness (Daft, 2010)(Lindgren 2012). This tendency can be argued to have influenced the BM literature e.g. Chesbrough (2007) suggests that BMs should be open i.e. OBM, which includes that businesses should utilize the resources of other businesses within their own BMs. It has been argued that until 2007, the BM literature was primarily regarding closed BMs (CBMs), whereas BMs were bound to the focal business, and thereby not open to other businesses (Lindgren, 2011). The CBM by Chesbrough (2007) was not deemed fit in the new global business environment, which requires openness and interfaces being able to comprehend interfacing with other businesses' BMs. Chesbrough (2007) further claims that CBMs delimit the potential value and effective use of the BMI. BMI, as mentioned in the introduction, refers to the reinvention of current BMs in order to create competitive advantages. Thus, Chesbrough's (2007) way of thinking of BMIs, as being open, has become the foundation of a development of a new network-based BM innovation concept. (Chesbrough, 2007) (Daft, 2010) (Lindgren, 2011). This foundation will also be adopted in order to define a GBM.

Globalization supported by ICT has increased the dynamic drive of the global environment and the requirement for agility (Friedman, 2007) (Lindgren, 2011). Osterwalder (2010) supports this by stating that BMs are becoming dynamic and that today's BMs may be outdated tomorrow. Furthermore, Lindgren (2011) suggests that new innovating BMs should serve as a platform for long termed and sustainable BMI. Any business model is a platform for BMI and thereby development of a multitude of BM's. The question is how do businesses "design" these BM's into GBM's?

4. BUSINESS CASES

In order to approach the combination of globalization and BMs to define a GBM, two cases are presented. The first case is based on a recent announcement from IBM, who is implementing a new BM in order to reinforce its innovation process. The second case is concerned with an already functioning BM which Boeing introduced in relation to the manufacturing of the 787 Dreamliner.

4.1. IBM – BELIQUID

IBM is an example of a business which has accepted and embraced the possibilities of an increasingly globalised world. IBM recently launched a new platform for a liquid business structure which assumedly is going to release some of the IBM business' fixed expenditures related to permanently employed staff and make the organization more agile (Henning, 2012). The new liquid business structure of IBM will emphasise the new dynamic global business environment which demands for high agility. The liquid program implemented by IBM is a new business platform emphasizing the opportunities of the global market, by allowing anyone to be a resource in the BM, and thus in the creation of value for IBM's clients (IBM, 2011). By doing so, IBM manages to avoid making long term relationships with suppliers and developers, which allows them to only utilize necessary resources when needed for a particular project, and to gain input to their BMI process from the entire global network. By implementing the related BMs, IBM has changed their network of partners to the so-called virtual agents of the BeLiquid program. As the name implies, the BeLiquid program is, opposite a set of long term partnerships, liquid. It is liquid in the sense that the virtual network partners are assigned to a single assignment at a time. (IBM, 2011)

4.2. THE 787 BOEING DREAMLINER

The famous Boeing 787 Dreamliner finally reached the market in September 2011, more than three years late (Gates & Allison, 2011). Even though both BMI budget and time schedules expired, Boeing took an initial step towards utilizing the global accessibility of resources in order to create an innovative BM ecosystem—the BM ecosystem related to the 787 Dreamliner (Boeing, 2008–2012). With 44 tier one suppliers from all over the world, (Boeing, 2008–2012) Boeing has taken outsourcing to a new level (Weitzman, 2011). Thus, the innovative 787 Dreamliner is considered as a paradigm shift (Weitzman, 2011), reliant on a GBM. In order to create BMI, Boeing divided the airplane into small subcomponents, in order to outsource each component to the global network of suppliers, who were most capable of producing the specific parts and services. The GBM can be argued to be a platform for developing other BMs. If Boeing had chosen to do parts of manufacturing in-house and to use only their current network suppliers, it was expected that the BMI had been limited by internal and partners' competences and the 787 Dreamliner had probably not been considered as a paradigm shift within airplanes. On the other hand, the Dreamliner might not have been three years late.

The demand for BMI in airplane manufacturing does not change as rapidly as it is the case for IBM's software BMI. Nevertheless, the Boeing case is still a good example of a business that is approaching BMI with an agile mindset. Airplane manufacturing is related to high fixed asset costs which easily make an airplane manufacturer inflexible and incapable of rapid BMI. Despite the unbalanced budget and the expired time schedule, Boeing is not left with heavy, costly and specific competences, and hypothetically Boeing would be capable of operating a new innovative BM by changing to a new range of suppliers of small subcomponents, who are capable of delivering the newest and most innovative technology.

Boeing created a unique BM ecosystem based on the availability of resources and competences from the global network of smaller and more specialized and less prejudiced network partners. The choice of the new BM ecosystem was, however, not unproblematic, and it showed to be a complex BM ecosystem to manage (Hiltzik, 2011). Nonetheless, important lessons were learned, not only for Boeing, but for organizations which are facing similar challenges from the increased globalization.

5. CONCEPTUALIZING - THE GLOBAL BUSINESS MODEL

In the following, globalization's impact on the BM concept enables us to discuss a definition of the GBM. Figure 1 summarizes and parallels globalization and BM evolution. In Table 1, it is illustrated how BMs have evolved from being limited to focus on the internal business to become an OBM platform, focusing on network-based BM and business model ecosystems. Thus, it is argued that BMs today are very much focused on utilizing the competences of the current 'delimited' networks. However, requirements for agility and opening up for the BMs are becoming more important in order to cope with the increasing complexity, and competition in BMI among others, caused among others by the fast development of ICT and the cloud-based BMIs (Neffics 2012). Furthermore, the life cycle of a BM and the first mover advantage period are shortened as a result of the ICT

Table 1 Globalization and Business Model development based and inspired by (Friedman2, 2005)

Time	'Size of the world'	Globalization Forces and Features	Business Models
Globalization 1.0 (1492–1800)	Size L – M	Globalization is driven by States and Governments	No focus on Business Models
Globalization 2.0 (1800–2000)	Size M - S	Globalization is driven by multinational Companies Motivated by -FALLING TRANSPORTATION COSTS	
		- TELECOMMUNICATION COSTS	Business Models are closed - Bound to the focal business (Lindgren, 2011)
Globalization 3.0 (2000-...) *2012*	Size S – XS	Globalization is driven by individuals (Empowered developed ICT) - INCREASINGLY NON-WESTERN - SPEED AND BREATH	Business Models are open (Chesbrough, 2007)Business Models are Network based Taran 2010 Businesses have one or more Business Models – the multi business model approach (Casadesus-Masanell 2010, Lindgren 2012) Business Models are realted to Business Model Ecosystems (Lindgren 2012)

driven globalization (Lindgren, 2011). Thus, requirements for speed in relation to BMI are increasing. This can be related to Friedman's (2009) and (Lindgren, 2011) view on globalization and BM's lifecycle, that the world is still shrinking, globalization and BMI are now driven by individuals and things (IOT European Commision 2013) and is moving towards a business landscape consisting of small entities, and everybody and everything is a business and member of one or more BMs (Friedman2, 2005) (Lindgren, 2011).

This raises interesting question—how can a Global Business Model be defined and what are the characteristics and challenges related to a Global Business Model?

Linking Friedman's (2005) statement that globalization is driven by individuals, to the definitions of globalization suggests that businesses and their related BMs must continuously reconsider if their networks are offering the best competences, resources and capabilities available to solve any issue. Partners in a network are, to some extent, bounded by current assets. The IBM and Boeing business cases illustrate how businesses are actually trying to utilize the BM opportunities of the global network and phase out the limitations of a delimited network.

Hence, it is suggested that smaller entities around the globe are more flexible and less prejudiced when solving issues. The ICT allows businesses and BMs to access almost the entire world and thereby link it to the global network. Individuals, the citizens of the globalized world, are already taking advantage of the opportunities emerging from globalization. Individuals and individual businesses are not limited by any competences, lack of competences or long termed partnership agreements when performing innovations in their businesses. Individuals, individual businesses and BMs can be as agile as they want. Businesses and BMs can attain knowledge and new required knowledge in few seconds by utilizing the global network—e.g. searching on Google, and through cloud networks or Facebook profiles, which means that individuals and businesses, potentially, are in constant contact with a network of many million active users (Facebook, 2012). Furthermore, individuals, businesses and related BMs are utilizing resources in the most efficient way, when doing business—utilizing the global network to ensure that they get the right performance, right costs, and right price at right time (Lindgren 2003). It can be argued that GBMs with success in the global market are expected also to be up-to-date related to the global markets context and demands. Businesses like IBM and Boeing are trying to embed the global and agile mindsets into their GBMs.

Most businesses are reliant on predefined and framed relationships with network partners. This can be a disadvantage in the increasingly globalized world and might slow down and limit the BMI process. A GBM utilises the opportunities which globalization offers. Hence, it is suggested that global businesses and their GBMs include the global network partners' building blocks in their BM's building blocks especially when innovating on the BM. The global network consists of creative and small entities, which are not limited by asset specificity, and which are the best within a specific field. The global network, based on Friedman's (2011) globalization view, is considered to be a huge accessible resource when innovating on BMs, which can solve BMI issues with high speed. It has been suggested, based on Lindgren (2011), that a GBM model should allow for sustainable BMI. The GBM is challenged with demand for agility, dynamic and should, hence, be able to take advantage of the global resource opportunities in order to interact within the complex and continuously changing global environment. It has been argued that the BM concept has shifted from being bounded to the focal business towards a more network-based business and BM perspective opening up for interaction with other businesses' BMs (Chesbrough, 2007) (Osterwalder2 & Pigneur, 2010).

The next step towards a GBM is to liquidize the BM concept. Osterwalder & Pigneur (2010) have argued that today's BMs are most likely to be outdated tomorrow. This implies for a change in our BM understanding from a static perception of a BM to a dynamic understanding of BMs. The GBMs are related to the idea of a sustainable BM. A sustainable BM must be agile to maintain its relevancy—it must be liquid rather than framed. In this paper, liquid, in relation to BMs, is referred to as the agility obtained by a variable and dynamic appearance of a BM, and regarded as a process of creating, capturing, delivering and consuming value (Neffics 2012)(Lindgren 2012). This means that the BMI is not limited to use only internal BM´s available in the closed partner network. This can be related to Masanell & Ricart's (2009) suggestion, that parts of a BM's building block must be kept variable in order to react to changes in the environment (Masanell & Ricart, 2009) —Global BM environment.

Osterwalder & Pigneur (2010), suggest that a BM can be divided into two sides; a value side and an efficiency side. We propose that the value side includes value creation, capturing, delivery and consumption related to BMs. The efficiency side according to Osterwalder2 & Pigneur 2010, can be expressed as the value creation part of the BM, including the activities and resources necessary to receive values. We propose that this should also include value capturing, delivery and consumptions. As argued previously, successful businesses and BMs' value sides are deemed to be rather agile today. This means that today's successful BMs should be capable of reaching various BM´s in the global market with incremental and radical changes related to all building blocks in the BM. Thus, the value side of today's BMs is deemed to be rather liquidized as business value, and their related BM´s value propositions are in a GBM made available to the entire world e.g. eBay, Amazon, Facebook, IBM and Boeing. In order to create a GBM, the efficiency side must as well utilize the full potential of ICT and opportunities related to the globalization, to take advantages of full potential of the global network and community in an agile and dynamic way. Based on this and, the IBM and Boeing cases, a GBM is therefore suggested to be identical to a liquidized BM. Therefore, the building blocks of a GBM in relation to the value flow (Neffics 2012) in a GBM can, based on Harold *et al.* (2008) and Lindgren (2011), come from everyone, everywhere, be everything and reached anytime.

Thus, a GBM is suggested to be equal to—a liquidized—flexible, agile, dynamic, sustainable BM that creates, captures, delivers and consume value, which allows for continuous GBM innovation by exploiting and using the entirety of the global networks business models.

Businesses have started to realize the real potential of the global network and of a more liquidized BM, in order to open up their BMI further. The IBM business case is exemplifying how a business is taking a step in the direction of GBMs. The Boeing case illustrates some of the obstacles which a GBM could face.

6. CONCLUSION

The BM concept is increasingly gaining acceptance within the business world. The concept has evolved within the increasingly globalized environment. Conversely, globalization is argued to be an ongoing and possibly unstoppable process. Thus, this paper has been concerned with the relation between the globalization and the BM concept. The research is driven by the assumed demand for a definition of the GBM and how to cope with the dynamic and complex environment in which businesses and their BMs have to interact, in order to guide future BMI.

The significance of the GBM is related to the huge possibilities that globalization is offering. On the other hand, the increased competition and the rapidly developing ICT have forced businesses and their BMs to become more agile and dynamic. In this context, the GBM is highly related to the resources and opportunities of the global network. The IBM and Boeing business cases have shown that steps are taken in the direction of embracing the global networks possibilities and global network partners' BMs. However, the Boeing case also illustrates that there are challenges in implementing the GBMs. Hence, the cases founded the basis for concluding that it is relevant to consider and define a GBM. From our findings, it can be concluded that a GBM is not a snapshot of a business's current or future way of doing business and BMI, but rather a liquidized and thereby sustainable rationale of how a business creates, captures, delivers, receives and consumes value, which allows for continuous BMI by exploiting and using the entirety of the global network.

This paper aims to encourage business managers to adapt their mindsets related to GBMs and BMI. It is suggested that a GBM is related to the full utilization of the opportunities available in the global network, in order to be agile, dynamic, flexible and sustainable. The focus of this paper is to contribute to the design process of the GBMs in order to be able to innovate the BMs in relation to the global market and compete in the increasingly globalized market. The GBMs engage the business and related BMs to an open minded approach of interacting with a huge number of accessible network partners. However, this raises significant challenges in relation to BM leadership and management.

7. FUTURE EXPECTED RESULTS/CONTRIBUTION

The study has enlightened a theoretical demand for a global and assumable liquid business model concept. The next step is to initiate a more thorough and empirical-based research to clarify the hypothesis, in order to be able to support, and test the statements and the proposals of GBM to propose specific managerial guidelines on how to innovate a GBM.

8. REFERENCES

[9] Abell, D. F., "Defining the Business: The Starting Point of Strategic Planning" New Jersey: Prentice-Hall, Inc., 1980.

[10] Boeing. (2008-2012). New Air Plane. Hentede 17. May 2012 fra Who is building: http://www.newairplane.com/787/whos_building/

[11] Chesbrough, H. (2007). Open Business Models How to Thrive in the New Innovation Landscape . Harvard Business School .

[12] Daft, R. L. (2010). Understanding the Theory and Design of Organisations. Vanderbilt: South-Western.

[13] Facebook. (2012). newroom.fb.com. Hentede 19. May 2012 fra Facebook: http://newsroom.fb.com/content/default.aspx?NewsAreaId=22

[14] Fielt, D. E. (2011, March 31 st). Outcomes. Retrieved April 13 th, 2011, from Smart Services CRC: http://www.smartservicescrc.com.au/PDF/Business%20Service%20Management%20Volume%203.pdf

[15] Friedman, T. (November 2007). The World is Flat 3.0. Lecture at Massachusetts Institute of Technology, MIT . Massachusetts, Massachusetts, USA: MIT.

[16] Friedman2, T. (2005). The World is Flat - A Brief History of the Twenty-first Century. Farrar, Straus & Giroux.

[17] Casadesus-Masanell Ramon and Joan Enric Ricart From Strategy to Business Models and onto Tactics Long Range Planning 43 (2010) 195e215

[18] Gates, D., & Allison, M. (26. September 2011). Boeing, ANA celebrate first 787 delivery. The Seattle Times .

[19] Ghemawat, P. (September 2001). Distance still matter. Harvard Business Review , s. 137-147.

[20] Henning, D. (11. February 2012). IBM launches new form of day-wage labour. Hentede 19. May 2012 fra World Socialist Web Site: http://www.wsws.org/articles/2012/feb2012/ibmc-f11.shtml

[21] Hiltzik, M. (15. February 2011). 787 Dreamliner teaches Boeing costly lesson on outsourcing. Los Angels Times .

[22] IOT European Commision 2013 http://www.internet-of-things.eu/

[23] Liquid Introduction and Overview (2011). [Film].

[24] Lambert, D. M., & Cooper, M. C. (2000). Issues in Supply Chain Management. Industrial Marketing Management 29 , 29 (1), 65-83.

[25] Lindgren 2003 Network Based High Speed Product Innovation. PhD Desserteation Center for Industrial production ISBN.

[26] Aalborg : Denmark., 2003.

[27] Lindgren, P. (2011). NEW global ICT-based business models. Aalborg, Denmark: River.

[28] Lindgren, P. (2012) Towards a Multi Business Model Innovation Model. / Lindgren, Peter; Jørgensen, Rasmus . Journal of Multi Business Model Innovation and Technology 1 edition River Publisher

[29] Masanell, R. C., & Ricart, J. E. (2009). From Strategy to Business Model and to Tactics. Hentede 17. May 2012 fra Harvard Business School: http://www.hbs.edu/research/pdf/10-036.pdf

[30] Neffics 2012

[31] Osterwalder, A., Pigneur, Y., & Tucci, C. (Vol. 15. May 2005). Clarifying Business Models. Communications of AIS .

[32] Osterwalder2, A., & Pigneur, Y. (2010). Business Model Generation. New Jersey: John Wiley & Sons, Inc.

[33] Peng, M. W. (2010). Global. Mason, OH: Sourth-Western.

[34] Porter, M. E. (No. 3. Vol. 79 2001). Strategy and the internet. Harvard Business Review , s. 62-79.

[35] Schenkar Oded and Betty Jane Punnett 2006 Handbook for International Management Research

[36] Sirkin, H. L. (2008). GLOBALITY: Competing with Everyone from Everywhere for Everything. New York: Hachette Book Group.

[37] Slepniov, D. (6. February 2012). Globalization. Unpacking Globalization . Aalborg, Denmark: http://ses.moodle.aau.dk/file.php/597/Files-lecture-01/G2012_06022011_session_1.pdf.

[38] Taran, Y. e. (2009). THEORY BUILDING - TOWARDS AN UNDERSTANDING OF BUSINESS MODELINNOVATION PROCESSES. Druid-Dime Academy 2009 PhD Conference. Aalborg, Denmark: Centre for Industrial Production.

[39] Vervest, P et al., 2005, Smart Business Networks Springer ISBN 3-540-22840-3

[40] Weitzman, H. (25. September 2011). Dreamliner lessons to shape Boeing direction. Financial Times .

[41] Wolf, M. (2004). What Liberal Globalization Means. I M. Wolf, Why Globalization Works (s. 13-19). New Haven and London: Yale University Press.

BIOGRAPHIES

Kristian Bonde Pedersen - He holds a B.Sc. in Global Business Engineering also from Aalborg University and is currently studying Master in Operation and Innovation Management. He is a research assistant at the Multi Business, Innovation and technology group. He has in his study worked close with several companies and focused on Business & Business Model Development and Innovation, Strategic Planning and Operationalization analytic and diagnostic, Operational Management – analytic and diagnostic, Supply Chain Management – analytic and diagnostic, Service and Motivation Management – analytic and diagnostic, Cultural understanding and communication.

Kristoffer Rose Svarre - is currently studying Master in Operation and Innovation Management, Center for Industrial Production at Aalborg University. He holds a B.Sc. in Global Business Engineering also from Aalborg University. His Bachelor project was about Business Model Innovation in a global business environment and he in his study he has focused on Supply Chain Management, Innovation and Change management among others. He have previously done 2 years of study in Software Engineering. He has, besides his studies, a job as a research assistant for the research group, M-BIT - Multi-Business, Innovation & Technology Group.

Dimitrij Slepniov – is an Assistant Professor PhD at Center for Industrial Production, Aalborg University and holds an MBA from Vilnius University, Lithuania and a M.Sc. from London School of Economics. He has (co-)authored numerous articles and books on subjects such as Globalization, supply chain management, global business development His current research interest is in Globalization and internationalization of business i.e. the typology and generic types of business globalization and how to internationalize them.

Peter Lindgren is Associate Professor of Innovation and New Business Development at the Center for Industrial Production, Aalborg University, Denmark. He holds B.Sc in Business Administration, M.Sc in Foreign Trade and Ph.D in Network-based High Speed Innovation. He has (co-)authored numerous articles and several books on subjects such as product development in network, electronic product development, new global business development, innovation management and leadership, and high speed innovation. His current research interest is in new global business models, i.e. the typology and generic types of business models and how to innovate them.

Conceptualizing strategic business model innovation leadership for business survival and business model innovation excellence

Peter Lindgren[1]

Maizura Ailin Abdullah[2,]

[1] Department of Mechanical and Manufacturing Engineering, Aalborg University
[2] The Royal Institute of Technology (KTH) in Stockholm, Sweden.

Received 12 February 2013; Accepted: 25 February 2013

Abstract

Too many businesses are being marginalized by blind "business model innovations (BMIs)" and simple "BMIs". As documented in previous research (Markides 2008, Lindgren 2012), most businesses perform BMIs at a reactive level i.e. perceiving what the market, customers and network partners might want rather than what they actually demand.

Few businesses have the ability to proactively lead BMIs and on a strategic level lead BMIs to something that fits the business's long term perspective (Hamel 2011). Apple, Ryanair, Facebook, Zappo are some businesses that have shown BMI Leadership (BMIL) in a proactive way — and more importantly, as some examples of first level BMIL.

The overall aim of the BMIL is to prevent businesses from being marginalized by the BMI and thereby to optimize the business's total BMI investment.

The literature research and case research we studied gave us some important inspiration, themes and baseline for conceptualizing BMIL and to formulate a framework proposing the BMIL strategy process. It also points to some of the requirements that should be taken into consideration and included to become successful via the BMI.

The paper focuses on the following research question:
- "How can businesses gain strategic advantage and learn business survival via BMIL?"

Keywords: Business model innovation, Business Model Innovation Leadership and management

1. INTRODUCTION - WHY BUSINESS MODEL INNOVATE?

Miller (1992) questions the notion of being "caught in the middle" or "caught in the innovation spiral". He can be claimed to say that there is a viable middle ground between business innovation strategies. Many businesses have entered a market with success as a niche player or a business focusing on other business model values or different business model values and cost structures (Ryanair, Zara Inditex, Starbuck, Yellow Tail) than established businesses in the industry and gradually expanded their businesses from there to become the leader of their business ecosystem and the BMI process in the business model ecosystem. In some cases, they have even disrupted the existing industry via BMIs. An up-to-date critique of generic innovation and BMI strategies and their limitations, including Porter, appears in Bowman, C. (2008) Generic strategies: a substitute for thinking?

The importance of innovation can however be traced back to the 1930's when Schumpeter first introduced the groundbreaking phenomena of disruptive innovation (Schumpeter 1934). Today, innovation is regarded as a fundamental condition for the survival of societies and businesses, whether they are big or small, even more so if they are small. Businesses are faced with the realities of perpetually-shortening business model life cycles and can no longer depend on short-term tactics, such as lowering costs and implementing minor differentiation or incremental improvements to their multitude of business models. Successful BMIs allow businesses to stay ahead of the competition in terms of cost, performance and development time to market. All these unseen advantages can translate to value to the business, customers and other stakeholders, allowing the business to ultimately stay at the front line of competition —but more importantly, in the frontline of the BMI process.

2. WHY DO BUSINESS NEED TO INNOVATE THEIR BUSINESS MODELS?

One possible answer would be: globalization. Globalization has, in more than one way, dissolved the boundaries between countries, economies, industries and organizations. It has brought about a ripple effect that affects all businesses in many ways. Technologies need to be upgraded, processes must be redesigned, communication has to be faster—all these, just to cope with the ever-changing needs of operations, customers, suppliers and global brands.

One after-effect of globalization is the usage of the Internet and "the cloud" in every day operations. The Internet has freed companies from the traditional ways of doing business, and maintaining relationships and networks. One simple example is BMI in the cloud. The application of "cloud-based BMI" has simply altered the relationships between customers, suppliers, value chains and the BMI processes. Information and knowledge travel faster, beyond measurable paradigms in the cloud. Customers and suppliers are now better-informed and well-educated about potential business models. This creates a power shift, placing the all stakeholders at an advantage in both a "TO BE" and a "AS IS BM" (Lindgren 2012).

The chain reaction goes further. Leading businesses understand the need for increased innovations in product innovation processes and the speed required to innovate products (Fine 1998, Lindgren 2003). In order to stay competitive and gain strategic advantages, businesses now have taken innovations one step higher by incorporating the BMIs (be it radical or incremental BMI) into their strategies. It has been proven that businesses that have introduced and implemented innovation strategies are better able to survive the competitive conditions, compared to companies that have not (Cooper, 2005). However, we still lack evidence that BMIs can prove the same.

All these mentioned above, when strategically combined and implemented, have the potential to improve work processes—product innovation timelines can become shorter, production costs can be lowered, while improving product quality significantly. The speed and efficiency with which innovations are diffused throughout an economy is thought to be critical in increasing productivity and economic growth. In addition, innovations are believed to possess the ability to prolong the survivability and competitiveness of businesses. However research have shown that the most innovative business and countries are not always the winners (O'Brian 2007, Ruchonen 2007) . The issue is to place businesses performing BMIs at an advantageous position in markets via BMI. An advantage possition L, compared to competitors and also other stakeholders.

And yet, innovation research initiatives for the past 50 years have only given us a fragmented understanding within the field of product innovation theory, service innovation theory and organizational theory. These research initiatives have provided us with some basic fundamentals of innovation and pointed out the complexity of innovation—put together, an opportunity to begin to study business model innovation leadership. BMIL attempts to place all the fragmented innovation components together and move our understanding of innovation further to a strategic level, i.e. from a management level to a leadership level. Such a topic reveals some new opportunities and challenges to innovation research, to the industry and the society.

BMIL for us is also about a continuous process of an integrated BMIL model which we propose as focusing on different levels of the BMI. For such a task, the management ideals are insufficient as BMIL requires vision, goals, strategy, sustained belief in BMI's success and a strong commitment to the BMI initiatives. It is about being able to form an integrated overview of the business innovation activities and concurrently "lead" the BMI in a strategic manner, in an ever changing world where a business has continuously to rethink its BMI conditions. It is not just about handling and managing an innovative product development project, rather it is about leading the business "BMI portfolio" strategically, which we shall now elaborate.

This article intends to introduce a brief overview of the available literature on innovation and leadership. Following this, it examines and defines the framework of the BMIL, consequently leading to our framework of the BMIL and thereby leading the BMI portfolio to a strategic advantageous position in the business.

3. THE NATURE OF BMI

BMI can briefly be described as something new, be it a value proposition, customer, value chain, competence, network, relation or a value formula (Lindgren 2012), that changes the basis of the business model —the way the business model is formulated and/or designed. The BMI can be something that is significantly improved, or "based on the results of new developments or new combinations of existing business model blocks."

BMI comes in many different varieties—change of one or more business model block(s), development of a new business model block, change of a business model's relations to other business models either internal or external the business and creation of a new business model ecosystem. BMI is regarded as something so uncertain that the best a business can do is to pour sufficient resources into it and then hope for the best. And, when the businesses are small- and medium-sized enterprises, they are even more dependent on successful outcomes of BMI as compared to large companies.

BMI can be classified into radical and incremental business model innovations (though the terms discontinuous vs. continuous innovation are also used interchangeably) (Balachandra 2000, Leifer 2002, Tidd 2003, Taran 2010). Radical business model innovation occurs very rarely, but the benefit and rewards are exceptionally high, whether they are financial or value-based rewards to the business or beneficial returns the society. Such rare occurrences involve a breakthrough in complexity, radicality and reach (Taran 2010). Radical BMI usually results in a large change in an existing "business model's core" or a new business model. Of course, the degree of radicality is scalable to the time of the BMI process potential measured related to three dimensions as shown in the model in Figure 1.

Incremental BMI, on the other hand, usually involves improvements and small changes in steps which are more progressive in nature. It occurs more frequently and is usually much easier for the business to carry out, simply because it is not as "foreign" and new as something which is a result of the radical BMI. The rule of thumb is that, the more common the business model innovation, the higher the potential of it to be successful. This is because it is more frequently based on tried and tested business models and BMI processes.

BMIs are a major challenge to many businesses today, as they suffer high failure rates within business models. This is due to many different reasons, among them being:

- a predominance of incremental business model innovation, which does not give long-lasting competitive advantages to the business;
- a high failure rate for BMI initiatives. Generally, only few ideas to business models usually reach their market potential, and are only successfully in the early stages of the BMI phase;
- a shorter business model life cycle for new business models, which means that up to 60 to 70% of new business models have to be re-developed within a short time after their introduction;

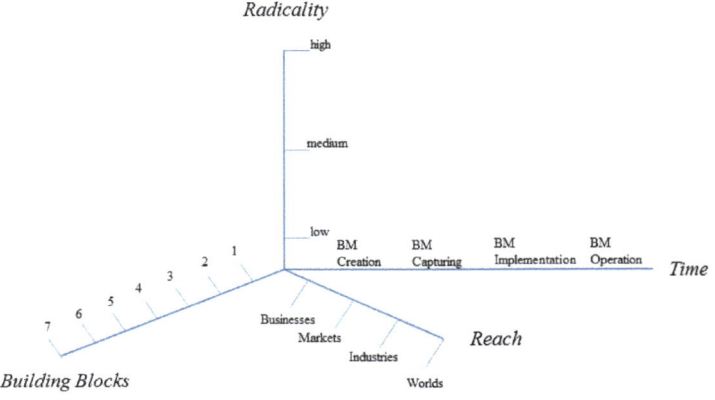

Figure 1. A three dimensional scale of BMI related to the BMI process.

- limited resources available for BMI (in SMEs) and inefficient BMI resource management in large businesses.

Perhaps, one of the biggest challenges in BMI research is measuring its outcome, efficiency and productivity, as will become evident in the upcoming section. It should be noted, however, that both radical and incremental BMIs are important to the survival of business, as both play important roles in the BMI strategy of business. We claim that the BMI success is not accidental; rather, it is a result of a combination of complex strategy thinking, capturing, dealings and actions, which when combined together, creates the basis of business sustainability. Some of these components of success will be elaborated in the next sections.

4. RESEARCH QUESTION

We studied the following research questions:
- How can businesses gain strategic advantages and business survival techniques through the BMIL?

In this context, the paper tries to conceptualize the fundamentals and basis of the BMIL. Second, the paper proposes a BMIL strategy process.

5. DESIGN/METHODOLOGY/APPROACH

The approach is a literature and qualitative research based on research carried out in the timeframe from 2002 to 2012, covering several national and international BMI projects (PUIN project, Newgibm project, ICI project, WIB project and Neffics EU-project funded by the European Commission).

6. EFFECTIVE AND EFFICIENT BMI

As mentioned previously, measuring the outcome and productivity of the BMI is a major challenge in this field. This is because each business around the world has its unique sets of criteria, goals and objectives with relation to its respective business agenda. These businesses then have different resources and varying levels of BMI skills and competences, not excluding knowledge repositories, which influence how business innovation is tackled and achieved in the respective business.

America has always had a history of rewarding creativity and churning out BMIs. However, these trends have been fast fading, according to several research projects (O'Brien 2005, Boston Consulting group 2012, INSEAD 2012, NAM 2012, The World Intellectual Property Organization, Cornell University 2013). As corporate and public nurturing of inventors and scientific research is diminishing in general, there is a need to rethink strategic BMIs not only in the US but also among other western countries. The United States and the EU have been faltering lately and the BMI's efficiency and effectiveness would pay a serious economic and intellectual penalty.

This downward trend is also observed by many institutions e.g. OECD, the National Academy of Sciences, European Commision (Horison 2020 and FP 7), that are very concerned that America and EU will lose their lead in BMIs, and worse still, will not be able to regain their lead. Both American and European politicians are expressing growing concerns on future economic competitiveness of western businesses. Both American and European business are proposing and even trying out the traditional cures for this downward trend, i.e. cost cutting, educational programs, creating a research and innovation culture, increase in federal funds for research and tax incentives, among others, in order to ensure their lead among the most business model innovative in the world.

During the late 2000's, Europe began to show great interest in BMIs. As a response to this community-wide interest, a succession of initiatives has been introduced to encourage and nurture the BMIs throughout Europe. In March 2012, the European Council launched the "Horizon 2020" which placed BMIs and SMEs in the centre of major policy efforts. Among its many commitments was to make Europe "the most innovative and dynamic knowledge-based economy in the world" by 2010. Such an ambitious effort required a rapid upgrade of EU's business model innovative capacity and an increase in the EU's BMI expenditure. This effort was in response to a report in 2010 which revealed that the EU and America lagged behind Asia in BMI investments.

The fourth Community Innovation Survey (CIS 2008) which prepared the "Innovation policy: updating the Union's approach in the context of the Lisbon strategy" is currently in the field, and CIS 2010 is still being planned and will be underway soon. In 2003, CIS stressed that the "innovation performance in the EU remained below levels recorded in the United States or Japan, and that a lack of innovation activity could be one of the key factors in explaining EU's underperformance in terms of productivity growth in recent years" (Luxembourg: Office for Official Publications of the European Communities 2004, 13).

The Asian region, with two-thirds of the world population, was advancing fast to challenge America's lead in research and innovation in the early 2000's (Silverthorne 2005) and continue to challenge America. Thus, globally, Asia can be considered one of the largest and fastest-growing investment locations for BMIs. This development continues to offer bigger and bigger potentials where advanced BMIs are concerned, and America and Europe cannot afford to ignore this direction of BMI growth that is occurring in Asia but also in Latin America and Africa.

It is no question that the Asian region is gearing up and looking forward to the next wave of BMIs (in the form of combined disciplines of BMIL). Japan, China and India are investing heavily within the new technologies of the BMI. With the rising usage of the cloud in business model practices, it is not difficult for businesses to create, capture, deliver and consume network-based business models across businesses, markets, industries and worlds.

But, is increased investment allocation and spending in the BMIs and creative BMI labs really the solution, or rather the ONLY solution, to success? Businesses, be them "large" or SMEs, in their bid to reach the finishing line of a successful BMI, might actually bypass the golden edge of BMI success without even realizing it.

Businesses have definitely benefited from the Internet's and cloud's ability to "send" work easily around the globe. But, this is not to say that this is without problems. Businesses face fragmentation in their BMI strategy and policies, which are usually in conflict with other business models internal to the business but also with customers, network partners and other stakeholders' BMI strategies and policies. Bureaucracy is difficult to penetrate, leading to imprecise fund allocation to the right BMI projects. More importantly, businesses, in general, are lacking the BMI culture that encourages creative and strategic BMI. The gap between industry and academia is too large, hindering free exchange of ideas and flow of information about BMIL, while education and research institutes do nothing to encourage an innovation culture of experimentation and hardly focus on how do we bring the business model ideas to the market and make them grow and benefit the businesses. The gap between academia and practitioners often lies in their differences of interests and values related to the BMI. The academia focuses upon improving and increasing innovations without thinking about the cost, efficiency and long-term benefits of the investment. The practitioners focus on their businesses and bottomline without opening up and releasing their real potentials in their business and business models. All these factors place tremendous pressure on society's management and investment in business development to create, capture, deliver and consume the successes of BMI initiatives and investment to gain long-term efficiency, effectiveness and learning. A new BMI agenda is needed both by academia, practitioners and society. An agenda that would be initially created as a network-based and would open the BMI platform in the clouds.

7. THE NATURE OF BMIL

BMIL is also a major challenge to businesses today. So far, studies on leadership related to the BMI have mainly attempted to provide guidance on how to define the leader's task and role from a management perspective while focusing on leadership competences and characteristics (Bryman 2004; Rooke 2005). These studies mainly concern discussions on individual leadership, as well as collective leadership (leadership by several managers in a group or as a team internal a business), but not leadership across different businesses and business models (both internal and external). In this context, "leadership in the clouds" is a concept where business managers have to carry out BMIL in the clouds together with other managers from different businesses. There are many studies on organizational leadership inside businesses, where the leadership of a business and various characteristics of leadership seen from a managerial, strategical and tactical perspective. When debating BMI, such studies mostly covered the management of single BMI projects, however, most often at a tactical level. In all these, leadership studies rarely focused on the strategic leadership of the BMI and further these studies do not take into consideration that the world and the BMI game has changed over the last 10–15 years taking the field of BMI to the clouds, to a network-based and open BMI-based context.

Dennis *et al.* emphasized years ago how the strategic BMIL phenomenon should look like, by presenting four main areas:

1. Strategic leadership as a **collective phenomenon**—where the strategic leadership of business models and BMIs requires contribution from more than a single individual business or business model.
2. Strategic leadership of BMI is a **processual phenomenon**—leaders need to mobilize other stakeholders in a system of interrelationships, rather than what they are.
3. Strategic leadership of BMI as a **dynamic phenomenon**—consists of the emergence, development, conduct, impact, performance and learning of management teams. This research area deals with the dynamic construction, deconstruction and reconstruction of BMIL roles over time according to the present context of the business, portfolio of business models, business model and

its building blocks together with the business model ecosystems that the business is operating in.

4. Strategic leadership of BMI as a **supra-organisational phenomenon**—BMIL roles and influences on such roles extend beyond focal business and business model boundaries. Here, collective BMIL must mobilize support and lead relationships, not only within the business, but also within its network to optimize the business performance of BMIs.

Porter (1985), Kotler (1994) and Malhotra (2000) have taken quite a different approach to BMIL, which they term **market leadership**. Malhotra defines market leadership as a business leading its position in a particular market or line of business and sees this as an optimum type of leadership. Kotler stresses the importance of having a defending market leadership. And, Porter proposes how to achieve market leadership, i.e. via cost leadership, differentiation or a focus strategy. However, none of these authors have mentioned achieving leadership via BMI i.e. BMIL.

Studies in the area of BMI have, quite surprisingly, hardly touched upon leading the market via strategic BMIs. Businesses that wish to ensure continued growth or competitiveness need to select one or more BMI champion(s), i.e. the right BMI leader who will have the BMIL skills, charisma and determination to lead the business portfolio of the BMI initiative. Taking into consideration the various theories discussed, we can ask the question "Is there a specific and distinctive form of BMIL?" And, considering the many different components of BMIs mentioned above, Is a different BMIL profile needed for today's BMI game? Our answer to this is a clear – Yes!

The significance of BMIs is widely acknowledged in a range of organizations, societies and in global competition. Thus, it is important for businesses to develop the ability to lead BMIs and to understand what BMIL is all about. The BMI is an ongoing, never-ending strategic process. Though there are available literature on managing innovation, they address mostly and mainly the issue of business survival. BMIL, however, has many more aspects to it than just management. Today, businesses have to lead themselves into the very core of the BMI process and make their businesses stay here via BMIs. Otherwise, they will suffer the role of being marginalized in the BMI process which, as we see, several western businesses both large and small business are doing today. That is one major reason to why western countries are losing businesses and jobs because they are not creating new and sustainable businesses.

8. THE FRAMEWORK OF BMIL

Many researchers have attempted to provide their notions on what aspects to consider when discussing innovation. For instance, how to define the product innovation development task (Roseneau 1983; Leifers 2002), how to characterize the field of product innovation development (Sanchez 1996: Child & Faulkner 1998; Goldman & Price 1998; Bohn & Lindgren 2003; Price 2005, (Bessant 1999), how to define the success criteria of product innovation development (Balachandra 1983; Boer 2002; Bohn & Lindgren 2004), the characteristics of the product innovation development model (Cooper 1986; Corso 2002; Cooper 2004; Bessant 1999; Christensen 2003), and identifying and choosing the right enablers for high-speed product innovation development (Fine 1998, Lindgren 2003).

Few have, in addition, tried to answer the questions of Why is leadership in BMI important to business companies? and, How are BMILs implemented in businesses?

Cooper (2005) has commenced research in the area by focusing on product leadership as a pathway to profitable BMI, presenting four points of performance of his Innovation Diamond: strategy of the business company, portfolio of BMI activities, process for new product development and the climate of the business company (how successful senior managers are in creating and fostering an business innovative culture). However, Cooper in his work only touches upon fragments of the complete pallet of BMIL opportunities.

Until now, studies have predominantly focused on the business's individual management of BMIs, particularly, as in Coopers case, the product innovation development which is just part of the value proposition building block and part of the BMI and the BMI process that starts from an idea and concept and ends when the business company prototype is ready to launch the business model to product in the market. Our notion of BMIL should not, however, be confused nor used interchangeably with the current ideas of BMI management of market leadership. It is an ideology on how to lead the different components and the business's BMI portfolio via the innovation leadership in a framework called BMIL, in order to achieve more strategic BMI success. Our definition of BMI success is strongly related to the leadership of "the core of the BMI process" via BMI which is strongly related to the long-term vision, mission, goals and strategies.

According to our research understanding, the management of business model product innovation today takes place mostly at an individual and tactical mid-management level. As a starting point and for a summary on how we visualize the difference between business model innovation management (BMIMA) and business model innovation leadership (BMIL), please refer to Table 1.

In the BMI context, we differentiate between BMIL and BMIMA. We consider BMIL as related to the strategic part of BMI and BMIMA as related to the tactical level of BMI (Lindgren 2003).

BMIL focusses on:
How to strategically and proactively lead the business portfolio of BMs and BMI activities into the core of the BMI process?

BMIMA focusses on:
How to tactically and proactively manage the business portfolio of BMs and BMI activities in the core of the BMI process?

BMIL´s overall aim is to bring the business into a better strategic BMI position and thereby into the core of the BMI process where the business has the opportunity to actively lead the game of BMI. The opposite position would leave the business with no opportunities to influence and change the BMI processes irrespective of whether the enterprise wants to join and change the BMI processes. This is not a preferable strategic position.

Up to this point, we claim that discussion and research on BMIs leave us with a rather fragmented picture of BMIL. In our mind, only one-seventh of the total BMIL has the potential. There seems to be hardly any research with specific focus on the combination of BMIL, the BMI portfolio and what is more, the strategic role that BMIL can play in businesses. The research until today on this topic is mainly related to organizational leadership dimension of BMIL, which is of course necessary, but quiet different to what we define as the BMIL.

Table 1 A basic summary of the differences between innovation management and innovation leadership.

Business Model Innovation Management (BMIMA)	Business Model Innovation Leadership (BMIL)
Short-term objectives relying on tactics	Long-term objectives built upon strategy and strategical objectives.
Internal focus with importance placed at the operational and implementation levels.	Internal focus stressing on operational and implementation levels PLUS external focus at the strategic level and integration with tactical level.
Success criteria based on cost, time, (superior) performance of BMI.	Success criteria based on continuous improvement and continuous innovation, learning, and innovation knowledge and capability development.
Prefers only minor performance improvements that can be provided by incremental BMIs.	Supports incremental innovation, but focus, at the same time, advocates riskier, radical innovation and BMI.
Depends mostly on organizational competences.	Depends on innovating organizational competences, and at the same time, encourages the exploration and exploitation of external sources of BMI competences, i.e. network partners' BMI competence.
Most of the time concentrates on one BMI project and process at a time.	Leads a portfolio of business model innovative projects and processes consisting of a balance of both incremental and radical BMI leadership process.
Stresses high speed BMI.	Stresses right speed for the BMI.
The business has an internal, almost short-sighted view of the BMI process as it does not follow through with the BMI process once the business model "leaves" the BMI phase and enters the business, market, industry and worlds. A transaction business model innovation approach. Elements of stakeholder feedback on BMI proposals are often regarded as after-BMI services.	The business has an overall view of the BMIL process and is located at the center of the BMIL process. This way, the business can strategically position itself in the market by exploiting and implementing the BMI.

A holistic, strategic concept of the BMIL is, therefore, still lacking. We find this rather peculiar, considering the importance that is being placed on the BMI and its strategy.

In this article, innovation leadership is more than product development or product leadership. A good starting point in defining our BMIL strategy process, therefore, should commence with identifying the strategic task of BMI, defining the context of BMI and defining the success criteria of BMI.

Table 2 Short-term and long-term success criteria for BMIL.

Short-term success criteria	Long-term success criteria
Time Cost (perceived and actual) Value (perceived and actual) Performance	Time, i.e. right speed, right cost, right performance Continuous improvement Continuous BMI Learning BMI efficiency BMI effectiveness Placed in the core of the BMI process Leading the BMI process

The model of BMIL strategy process is shown in Figure 2. The figure starts with the analyzing and choosing process among the different types of strategic types of BMIs that businesses can and should follow in order to accomplish both short-term and long-term success of business model innovation, finally ending up with BMI strategy implementation, control, adjustment and correction.

Our proposed framework for business model innovation leadership introduces eight main focus areas to consider.

1. The building block dimension.
2. The business model dimension.

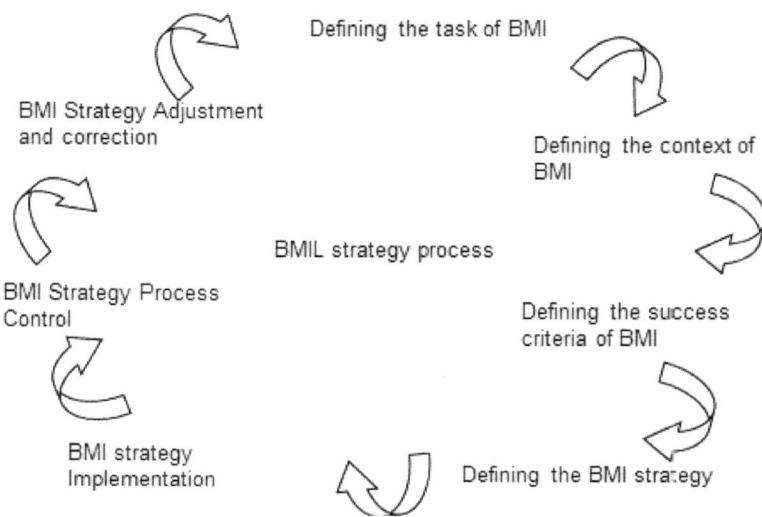

Figure 2 The BMIL strategy process.

3. The BMI dimension related to the creative part (both on AS IS and TO BE BMs) of BMI.
4. The BMI dimension related to the capturing, delivering and consuming part which we call "act and do" part of BMI.
5. The BMIL dimension—different viewpoints of BMI.
6. The portfolio dimension of a business, in this case, the integration and synergy between different business models and BMI projects on different models in the business.
7. The BMIL Strategy dimension—a business BMI strategy related to different phases in a BMI process.
8. The BMI strategy related to different business model ecosystems.

These eight BMIL areas have to be led individually, as well as together.

9. FINDINGS AND DISCUSSION

Perhaps, one way of visualizing the effectiveness and efficiency of a BMIL is to implement it in a innovation leadership portfolio and canvas.

Many CEOs we studied believed in pouring a large part of their resources into just one area of BMIL i.e. value proposition innovation leadership with high-investment, high-risk projects, with a belief that this one project is their only "golden egg" which will provide them with a jackpot of returns. Achieving success this way can often be attributed to pure luck. Such projects usually involve radical innovation and new knowledge. What happens is that the project would require sophisticated knowledge and thinking and it supersedes the project innovation timeframe. When this happens, the project usually requires further injection of investment, year after year, draining the available innovation resources from the business. Eventually, the BMI project is deemed unfruitful and the management is forced to pull the plug on the project, at the expense of many years of

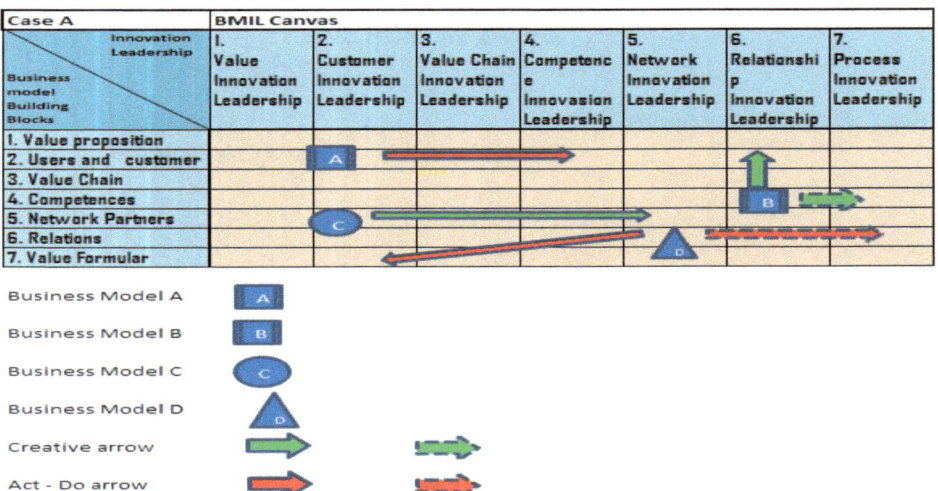

Figure 3 "TO BE" BM and "AS IS" BM in the BMIL Canvas – a case.

research and development and business resources that could have been deployed to other BMI areas and projects.

Just as investments can be made in portfolio style, so can a business's investment for business innovation projects—which in this case, we shall label as a "BMI portfolio". The idea is to diversify a business's resources, selecting a mix of business innovation projects to invest in and run, thus spreading the business innovation investment risk among different types of business innovation projects, according to the following factors.

It is believed that in the field of finance, portfolio analysis considers investors to be risk averse meaning that given two assets that offer the same expected return, investors will prefer the less risky one. Likewise, in an innovation portfolio, we assume that managers usually prefer to invest in innovation projects which are less riskier and require fewer resources. And, though they have a higher success rate, they provide lower returns, which do not contribute much to a business's profit margin (Leifers 2002).

A business's innovation portfolio in this sense should ideally consist of several carefully and strategical elected BMI projects of varying degrees of BMI initiatives. These projects should, at the same time, reflect the varying amounts of investments (whether financial, manpower or physical) needed to drive the respective BMI projects. The amount of investment assigned to each BMI portfolio item should depend on several factors, among them being:

- The type of BMI project being developed—whether it is a known BMI project that is available to the customers or a completely new business model that has not been seen before by market, industry or the world.
- The type of BMI that shall be utilized—whether incremental or radical innovation. We should also mention that this factor is closely related to the earlier factor of whether it is a known or new business model.
- The perceived receptivity of the business model in the market—customer and network partners' perceptions can help in the cases of known business models using incremental innovation. However, in the case of completely new business models being developed using radical business model innovation, there are available opinions that agree that e.g. just market research cannot satisfy this one factor (Christensen 1997, Drucker 1985, Cooper 2005).
- The expected timeline of the BMI project has the disadvantage of high development costs, long-term development periods and uncertain success rates.

The final factor that should be considered, just as in any BMI portfolio, is the risk factor. The risk assigned to each BMI portfolio component, whether it is a network-based BMI project, a BMI process project or a high-risk BMI project (with a high probability of failing, which also means that it probably requires more investment, which means that if it is successful it will result in higher returns, and if it fails, it will conversely result in higher losses) or a low-risk project (with a high probably of success, but demands low investment, and does not result in high returns when it is successful).

Having these factors in mind for a BMI portfolio will make a drawing of the strategic BMI leadership map of the business innovation leadership model more effective, in other words, which component deserves more attention and resources, at which stage of e.g.

BMI phase and "the business model lifecycle" should we pay more attention and investment.

Finding the perfect balance and combination of our proposed BMIL components is our interpretation of how to begin to lead BMI more strategically and bring business into BMIL.

10. CONCLUSION

BMI is not a new concept in business. However, the idea of continuous and sustainable BMI is fairly new. Businesses must learn to identify the opportunities of BMIL, react faster to changes in the field of BMIL and produce new BMI roads of BMIL faster, while balancing value, time, and cost together with continues innovation, continues improvement and learning. It is partly because of such dramatic changes in the game of the BMIs that BMIL has become a crucial and necessary ingredient for business growth and survival.

This, thus, triggers an urgent need for a new and improved thinking about leadership of BMIs. This is because business survival depends on the ability of its leaders to develop creative responses to the different types of challenges facing the BMI portfolio and BMIL. For successful implementation, top management must undertake a holistic approach to implementation and align the business innovations strategically, effectively and efficiently, both from a short-term and especially a long-term perspective.

BMIL is about developing and implementing a superior capability to innovate business models. It forces both an "outside" and an inside" look at the BMIL processes. This "outside look" manifests itself in the ability to integrate the BMI entities and processes external to the business (hence, outsiders) and integrating them into the business, making them part of the business innovation leadership strategy. The "inside look" manifests itself in the ability to integrate the BMI entities and processes internal to the business (hence, insiders) and integrating them into the different BMI projects and processes, making them part of the business innovation leadership culture. Fulfilling this BMIL vision, related goals and strategy brings the businesses into a position of leading the BMI process—a proactive BMIL strategy opposite to a reactive BMIL strategy, thereby into a strategy advantage position via strategic BMI.

This article introduced a slightly different approach to using the BMI in order to enable a business to achieve superior strategic reach and BMI position. Focusing on factors internal to the business (such as the business model blocks, business models, business model portfolio innovation process of the business), as well as external factors such as the business model ecosystem innovation process and hereunder the network-based BMI process. The aim of this process of BMI implementing the BMIL in the business is to place the business in a more central and strategic BMI position i.e.in the core of the BMI process. This allows the business to have an overall view of the BMI process, influence the BMI process and react earlier to forthcoming BMI processes. In this way, the business can strategically position itself as the leader of the market and become the leader of the BMI process by exploiting and implementing innovation.

Our proposal to the concept of BMIL is, therefore, different to what has already been said. The differences are mainly related to a move from tactical management of BMI to a more strategic BMI focusing on the strategical advantage in the business model ecosystems via BMIL.

Further, our concept of BMIL is more holistic involving seven dimensions of strategic BMIs i.e. value innovation leadership, customer innovation leadership, value chain innovation leadership, competence innovation leadership, network innovation leadership, relations innovation leadership and process innovation leadership. This forms the BMIL "umbrella" and potential that has to be orchestrated.

11. FUTURE EXPECTED RESULTS/CONTRIBUTION

We expect, in future research, to find more tools and methods for the BMIL. We expect that these will influence the possibilities for implementing BMIL.

12. REFERENCES

[42] Abell, D. F.., "Defining the Business: The Starting Point of Strategic Planning" New Jersey: Prentice-Hall, Inc., 1980.

[43] Balachandra, R. and Friar, J. H., "Managing New Product Development Processes the Right Way," 1 (1999) 33-43, Information Knowledge Systems Management, IOS Press

[44] "Factors for Success in R&D Projects and New Product Innovation: A Contextual Framework," (August 1997) IEE Transactions on Engineering Management, Vol. 44, No. 3

[45] Bessant, John. Challenges in Innovation Management. Brighton: Centre for Research in Innovation Management, 1999

[46] Bohn, K & Lindgren, P, 2002, 'Right Speed in Network Based Product Development and the Relationship to Learning, CIM and CI', CINet, Helsinki.

[47] Boer, H and Gertsen F From Continuous Improvement to Continuous Innovation: A (retro)(per)spective, International Journal of Technology Management.

[48] Bryman, A. "Qualitative research on Leadership: A critical but appreciative review." The Leadership Quarterly, 2004.

[49] Bowman, C. (2008)

[50] The Global Innovation Index by The Boston Consulting Group 2012 is a global index developed as a one off exercise back in 2009. It measures the level of innovation of a country. It is produced jointly by The Boston Consulting Group (BCG), the National Association of Manufacturers (NAM), and The Manufacturing Institute (MI), the NAM's nonpartisan research affiliate. NAM described it as the "largest and most comprehensive global index of its kind".

[51] The Global Innovation Index by INSEAD,

[52] The World Intellectual Property Organization 2013

[53] Child, J & Faulkner D,1998, 'Strategies of Co-operation – Managing Alliances, Networks, and Joint Ventures', Oxford University Press, Oxford.

[54] Casadesus-Masanell Ramon and Joan Enric Ricart From Strategy to Business Models and onto Tactics Long Range Planning 43 (2010) 195e215

[55] Cornell University 2012

[56] Chesbrough, H. (2003). The Era of Open Innovation. MIT Sloan Management Review, 44 (3), 1-9.

[57] Chesbrough, Henry (2006), Open Business Models:How to Thrive in the New Innovation Landscape, Boston: Harvard Business School Press.

[58] Chesborough, H. (2007) Open business models. How to thrive in the new innovation landscape, Boston: Harvard Business School.

[59] Chesbrough 2011 Keynote Speech at the Oslo innovation week October 2011

[60] Christensen Clayton and M. Johnson, What are Business Models, and How are they Built? Harvard Business School Note 9-610-019(2009)

[61] Christensen, Clayton M. The innovator's dilemma: when news technology cause great firms to fail. Boston: Harvard Business School Press, 1997.

[62] Cooper, Robert G. Product Leadership: Pathways to Profitable Innovation, 2nd ed. New York: Basic Books, 2005.

[63] Cooper, R, 1993 'Winning at New Products' Addison-Wesley Publishing Company ISBN 0-201-56381-91993

[64] Fine, C.H. Clockspeed, Perseus Book, 1998

[65] Francis, D. and Bessant, J. (March 2005). Technovation. "Targeting innovation and implications for capability development" Volume 25, Issue 3, March 2007, pp. 171-183

[66] Goldman, Nagel & Price, 1998, 'Agile Competitors and Virtual Organisations', Van Nostrand Reinhold, New York.

[67] Hayward, Bob M. "Innovation in Asia/Pacific and Japan Becoming World-Class." Gartner, March 10, 2006. Downloaded from http://www.gartner.com/DisplayDocument?id=489655&ref=g_sitelink on 25th December 2007.

[68] Horizon 2020 and FP 7 European Commision http://ec.europa.eu/research/fp7/index_en.cfm

[69] Johnson M.W., Christensen, M.C. and Kagermann, H. (2008) Reinventing your business model, Harvard Business Review, vol. 86 No. 12, pp. 50-59

[70] Johnson M., C. Christensen and H. Kagermann, Reinventing your business model, Harvard Business Review 86(12) (2008);

[71] Jia Hepeng. "China needs to encourage 'bottom-up' innovation." Science and Development Network, October 12, 2007.

[72] Kotler, Philip, Marketing Management: Analysis, Planning, Implementation and Control, (town unknown): Prentice-Hall, 1994

[73] Leifer, R. Critical Factors Predicting Radial Innovation Success. New York: Rensselaer Polytechnic Institute, December 2002.

[74] Lindgren, P. "Network Based High Speed Product Innovation" (ISBN 87-91200-15-6) PhD diss, Center for Industrial Production, Aalborg University, 2003.

[75] Lindgren P., 2012 Business Model Innovation Leadership: How Do SME's Strategically Lead Business Model Innovation? I: International Journal of Business and Management, Vol. 7, Nr. 14, 07.2012, s. 53-75.

[76] Markides C., Game-Changing Strategies: How to Create New Market Space in Established Industries by Breaking the Rules, Jossey-Bass, San Francisco (2008).

[77] Magretta, J. (2002) Why business models matter? Harvard Business Review, Vol. 80, No. 5, pp. 86-92.

[78] Malhotra 2000 (other details unknown).

[79] Miller (1992)

[80] Neffics 2011/2012 Baseline analysis of Business Values D 3.1., D 3.2. and Business Model innovation leadership D 4.1., D 4.2., D.4.3 the Neffics project 2012 www.neffics.eu

[81] O'Brien, Timothy L. "Are U.S. Innovators Losing Their Competitive Edge?" International Herald Tribune, November 13, 2005, Technology & Media section. Downloaded from http://www.iht.com/articles/2005/11/14/business/invent.php on 25th December, 2007.

[82] Organisation for Economic Co-operation and Development (OECD). "Measuring Innovation in OECD and non-OECD countries." (ISBN 0-7969-2062-1) Human Sciences Research Council, Cape Town, South Africa, 2006.

[83] Osterwalder, A, Y. Pigneur and L.C. Tucci (2004), Clarifying business models: Origins, present, and future of the concept, Communications of AIS, No. 16, pp. 1-25.

[84] Osterwalder et all 2010 Business Model Generation

[85] Padma, T. V. "India 'lagging behind' in innovation race." Science and Development Network, October 15, 2007.

[86] Porter, Michael E., Competitive Advantage. New York: The Free Press, 1985.

[87] Porter, M. E. (2011), Creating Shared Value: Redefining Capitalism and the Role of the Corporation in Society, Harvard Business Review,

[88] Rosenau, M.D., Managing the Development of the New Products, ITP, pp. 39-41., 1993

[89] Rooke, D. Harvard Business Review, 2005 (other details unknown).

[90] Ruchonen Juha (2007) Victa – Virtual ICT Accelerator Technology Review 219/2007 Teces, Finland

[91] Sanchez, R 2000b, 'Product, Process, and Knowledge Architectures in Organizational Competence', Research Working Paper, Oxford University Press, 2000-11.

[92] Sanchez, R 1996a, 'Strategic Product Creation: Managing New Interactions of Technology, Markets and Organizations', European Management Journal Vol 14. No 2, pp 121-138.

[93] Silverthorne, Sean. "The Rise of Innovation in Asia." Harvard Business School, March 7, 2005. Downloaded from http://hbswk.hbs.edu/item/4676.html on 25th December, 2007.

[94] Taran, Yariv Rethinking it All : Overcoming Obstacles to Business Model Innovation. Aalborg : Center for Industrial Production, Aalborg University, 2011. 193 p. Publication: Research › Ph.d. thesis

[95] Tidd, J., Bessant, J. & Pavitt, K., Managing Innovation: Integrating Technological, Market and Organizational Change, 3rd ed. New Jersey: John Wiley & Sons Ltd., 2003.

[96] Ulrich, KT & Eppinger, Product Design and Development. 2nd ed, San Diego: Irwin McGraw-Hill, 2000.

[97] Wind, Yuromoram. "A New Procedure for Concept Evaluation." Journal of Marketing (October 1973): 2-11.

[98] Scozzi 2012 Different Practice to implement Open Innovation Ifkad conference 2012

[99] Shafer S. M., H. J. Smith and J. C. Linder, The Power of Business Models, Business

[100] Horizons 48, 199e207 (2005).

[101] Taran, Y., Boer, H., & Lindgren, P. (2009). Theory Building - Towards an Understanding of Business Model Innovation Processes. Aalborg University, Centre for Industrial Production, Denmark.

[102] Teece David J. (2011) Business Models, Business Strategy and Innovation Long Range Planning 43/2-3 April/May 2010

[103] Ulrich, KT & Eppinger, SD 2000, 'Product Design and Development", 2nd edition, Irwin McGraw-Hill.

[104] X. Lecoq, B. Demil and V. Warnier, Le Business Model, un Outil d'Analyse Strate´gique, L'Expansion Management Review 123, 50e59 (2006).´

[105] Zott, C., and Amitt, R., and Mazza, L. (2010) The Business Model: Theoretical roots, recent developments, and future research. Madrid, Spain: IESE Business Shoolo, Ducument Number)

BIOGRAPHIES

Peter Lindgren is Associate Professor of Innovation and New Business Development at the Center for Industrial Production, Aalborg University, Denmark. He holds B.Sc in Business Administration, M.Sc in Foreign Trade and Ph.D in Network-based High Speed Innovation. He has (co-)authored numerous articles and several books on subjects such as product development in network, electronic product development, new global business development, innovation management and leadership, and high speed innovation. His current research interest is in new global business models, i.e. the typology and generic types of business models and how to innovate them.

Maizura Ailin Abdullah is a PhD researcher at the Royal Institute of Technology (KTH) in Stockholm, Sweden. She is attached to the Integrated Product Development group of the Department of Machine Design, under the School of Industrial Engineering of Management. At KTH, her research areas include Open Innovation, Business Models and Networking within Open Innovation scenarios.

Online Manuscript Submission

The link for submission is: www.riverpublishers.com/journal

Authors and reviewers can easily set up an account and log in to submit or review papers.

Submission formats for manuscripts: LaTeX, Word, WordPerfect, RTF, TXT.
Submission formats for figures: EPS, TIFF, GIF, JPEG, PPT and Postscript.

LaTeX

For submission in LaTeX, River Publishers has developed a River stylefile, which can be downloaded from http://riverpublishers.com/river publishers/authors.php

Guidelines for Manuscripts

Please use the Authors' Guidelines for the preparation of manuscripts, which can be downloaded from http://riverpublishers.com/river publishers/authors.php

In case of difficulties while submitting or other inquiries, please get in touch with us by clicking CONTACT on the journal's site or sending an e-mail to: info@riverpublishers.com

www.ingramcontent.com/pod-product-compliance
Ingram Content Group UK Ltd.
Pitfield, Milton Keynes, MK11 3LW, UK
UKHW021045200426
11947UKWH00041B/778